Nanotechnology and Ethical Governance
in the European Union and China

Sally Dalton-Brown

Nanotechnology and Ethical Governance in the European Union and China

Towards a Global Approach for Science and Technology

 Springer

Sally Dalton-Brown
Trinity College
University of Melbourne
Melbourne, Australia

ISBN 978-3-319-36630-2   ISBN 978-3-319-18233-9   (eBook)
DOI 10.1007/978-3-319-18233-9

Springer Cham Heidelberg New York Dordrecht London
© Springer International Publishing Switzerland 2015
Softcover reprint of the hardcover 1st edition 2015

Printed on acid-free paper

Springer International Publishing AG Switzerland is part of Springer Science+Business Media (www.springer.com)

Acknowledgments

This book has been partly written in the UK and partly in Australia (as well as on a few long plane journeys in between). I owe many people my thanks in both places and some beyond.

First of all I would like to thank the incomparable Empress Doris for her calm perspicacity and endless patience and for her great friendship. Thanks also to Miltos Ladikas for access to GEST reports, advice on pTA, and particularly for including me on his GEST trip to Beijing. I am grateful also to Ma Ying from CASTED for sending me various nanoarticles, and arranging a nanotechnology panel in Beijing, and to Camilo Fautz at ITAS (KIT) for sending me some useful articles on EU regulation. Thanks to Roger Chenells and Kate Chatfield, study buddies extraordinaire, and to Tim Thornton.

I also thank the University of Central Lancashire for providing me with the funding that made this possible and to Cathy Lennon for helping me with my many bureaucratic incompetencies, amongst other things.

And last but not least, thanks to you-know-who for the support, including endless cups of tea; by my reckoning this project took nearly a thousand. I say nothing of any other beverages that may have been harmed during the making of this book.

Contents

Abbreviations (Acronyms)

BASF	Chemical corporation, Germany
CASTED	Chinese Academy of Science and Technology for Development
CENARIOS	Certifiable Nanospecific Risk Management and Monitoring System
COMEST	UNESCO Commission on the Ethics of Scientific Knowledge and Technology
DBT	Danish Board of Technology (*Teknologirådet*)
EGE	European Commission's Group on Ethics in Science and Technology
ETAG	European Technology Assessment Group
EGAIS	Ethical Governance of Emerging Technologies project funded by the EU 7th Framework Program (*Science in Society*)
ELSI	Ethical, legal and social implications analysis
EU-CoC	European Commission's Code of Conduct for Responsible Nanosciences and Nanotechnologies Research
GEST	EU 7th framework Global Ethics in Science and Technology project
GHS	Globally Harmonised Scheme (for classification and labelling of substances)
GM	Genetically Modified
HLEG	European Commission's High Level Expert Group on Key Enabling Technologies ICON
ICON	International Council on Nanotechnology
IG-DHS	Interest Group – retail trade (Switzerland), a group of the six largest Swiss retailers: Migros, Coop, Denner, Manor, Vogele and Valora
IPM	Institute of Policy and Management, part of the Chinese Academy of Sciences
IRGC	International Council on Risk Governance

ISO	International Standards Organization
MOST	Chinese Ministry of Science and Technology
NGO	Non-governmental organization
NELSI	Nanotechnology ethical, legal and social implications analysis
pTA	Participatory Technology Assessment
PTA	Parliamentary Technology Assessment
REACH	Registration, Evaluation, Authorisation and Restriction of Chemical Substances regulatory framework
S&T	Science and Technology
SCER	Euro European Commission's Scientific Committee on Health and Environmental Risks
SCENHIR	European Commission's Scientific Committee on Emerging and Newly Identified Health Risks
Oviedo Charter	EU Convention for the Protection of Human Rights and Dignity of the Human Being with regard to the Application of Biology and Medicine: Convention on Human Rights and Biomedicine
WPMN	OECD Working Party on Manufactured Nanomaterials
WPN	OECD Working Party on Nanotechnology

Chapter 1
Introductory Context

Now nanotechnology had made nearly anything possible, and so the cultural role in deciding what should be done with it had become far more important than imagining what could be done with it. (Stephenson 1995)

Our stubborn egocentricities lead us to believe that the world must be run according to our own templates; that all would be well with the world if only it were populated with people like us. But our dehumanised apprehension of the familiar points directly to the resolution. Only when we come to understand, or even just acknowledge, each other's complexities can comprehension begin. Only when we apprehend each other as human…will the world begin to make sense. (Aly 2007)

Abstract Nanotechnology serves as a useful test case when asking the broad question of how the world might approach the issue of the societal impact of new technologies on a global basis. Nanotechnology is a key site for efforts to engage members of the public in dialogue about emerging technologies; East and West however differ markedly in terms of the degree to which such public involvement is facilitated.

Keywords Public engagement • Nanotechnology • Global ethics

Nanotechnology has been described as equivalent to the development that kick-started the industrial revolution (Arnall 2003); as a potential road to salvation for developing counties dealing with issues of energy production, agricultural productivity enhancement, and water treatment (Schummer 2007); but also as a Faustian bargain that may yet lead to global destruction (Joy 2000). Although at present nanotechnology is used in a variety of products, ranging from skin creams and food packaging to longer-lasting tennis balls and warmer socks, its potential is as yet undefined (solar power for all! A cure for cancer! Nanobots!), leading to both utopian hype and dystopian dread. Is it a form of miraculous engineering, a universal nostrum, or a terrifyingly vague science that may ultimately reduce the universe to 'grey goo' (Drexler 1996)? Or is it simply a new line of research that has already proved disappointingly unable to live up to its potential (Illuminator and Miller 2010)?

© Springer International Publishing Switzerland 2015
S. Dalton-Brown, *Nanotechnology and Ethical Governance in the European Union and China*, DOI 10.1007/978-3-319-18233-9_1

1

The facts are that nanotechnology will affect our lives (as indeed it already does through products available on the market), and that surveys show it to be poorly understood by the public in terms of its risks and benefits.[1]

An emerging science, nanotechnology provides an opportunity for the public to work alongside researchers, corporations, and governments to attempt ethical implementation. It provides a test case for how the world might act given a technology that could conceivably transform the global economy and provide solutions to major global issues such as pollution while potentially creating new environmental and health risks. Nanotechnology, in particular, has been presented as a key site for experimenting with novel forms of so-called upstream engagement or efforts to engage members of the public in dialogue about emerging technologies (ENSAA 2011). It provides a context for the emergence of a new risk governance paradigm, in terms of which local societies' political cultures and risk perceptions become significant factors in risk assessment and governance, with conventional top-down/authoritative decision-making models now taking more notice of public risk discourse (Chou 2007).

How might the world act together to inculcate ethical concerns into nanopolicy as it is formulated across the globe? This broad question may seem both idealistic and futile, yet it is only a restatement of a much older question about how to ensure that new technologies affect us only with our consent. This question raises several others:

1. What are the particular and/or new ethical concerns that might be raised by a new technology such as nano?
2. How do we work together globally to agree on such concerns?
3. And who is 'we'? What skills and processes are required for such global participation?

Given these three questions, encompassing a discussion of nanoethical issues, another of global cooperation, and also of skills and processes, this study firstly offers a comparative understanding of the environments within which nanotechnology is 'regulated' and discussed within China and the EU. Given these differences, this book then looks at how global ethics might work, offering a proceduralist model, one of pTA (participatory technology assessment). Thus the primary aim of this study is, namely, to provide an expanded pTA model, given the premise that global ethics is best done through global agents in pTA.

Habermasian thought allows the expansion of current pTA models to predicate how agents might achieve consensus – perhaps even global consensus (as Habermas's 'universalisation principle' suggests).

Nanotechnology as a globally significant new technology provides a useful case study against which to examine how pTA (e.g. nanofora) might work within and across differing cultures such as those of the EU and China. Given nanotechnology's significance in both regions as central to economic growth, agreement on the ethical impact on society of such a potentially radical new technology would seem to be of considerable importance.

---

[1] A 2010 report, for example, noted that 'only 45 % of Europeans say they have heard of nanotechnology'. See Gaskell et al. (2010).

1.1   New Concerns?

The first question is a very broad one, encompassing issues of military usage, distributive justice (given that nanotechnology could have vast application in developing countries), and risk management of toxicity, as well as more futuristic scenarios to do with human enhancement. No wonder that the range of potential nanotechnology applications has led to debates about public involvement in decision-making.

Science journalist Philip Ball succinctly stated the problem in his article on why 2003 was 'the year that society (in the UK that is) woke up to nanotechnology and got very alarmed'. Following the publication of Michael Crichton's *Prey* in late 2002 and the start of public debates in the UK media on the dangers of nanotechnology, Ball (2003) suggested the following ethical questions should be asked:

> Is nanotechnology primarily about wealth creation, or improving our quality of life, or something else? Who is developing it, and why? With what responsibility, justification, and accountability? Who deals with potential problems, and how? Is there, and should there be, a public mandate for it? The scientific community has no excuse for ignoring such questions.

Ethics and socio-ethical analysis have increasingly become an integral part of the assessment of any new technology and its applications (Berne 2004). Some countries, for example, the Netherlands, have instigated national programmes dedicated to ethics in science and technology (S&T) or have established academic institutes to conduct research into the ethical implications of new technologies. In China, ethical, legal, and social implications (ELSI) analysis is becoming a standard for evaluating new S&T developments.

Global variation in ELSI approaches to S&T development is however rather too large a topic to be useful; this study attempts to narrow down the notion of 'global', by 'West' referring chiefly to the EU, and by 'East', principally to China.

Whereas the EU has a history of attempting communal social policy across its member states (currently 27, with another 7 in candidature), the East has tended to form economic policy clusters, although the Association of Southeast Asian Nations, or ASEAN (currently ten members), which also looks at social policy, can be noted. There is no common approach towards an Asian policy on nanotechnology, as there is in the EU, and China has thus been selected for two reasons. China is one of the two largest Eastern 'players' in nanotechnology S&T (the other being Japan). China offers an interesting example of a country that fully perceives the benefits of a new technology, including its potential in the field of convergence technology (bio-/cogno-nano), while arguably having a less regulatory approach to such economically promising science than the EU:

> Like no other country China understood that to win the race depends on finished products through Nano-Bio-Cogno-Info convergence.... Second, but maybe even more important, there are no ethical restrictions or social controversy on developing and using nanotechnologies for new products and systems. (Nanotechnology Now 2007)

China's massive investment in nanotechnology research indicates the attempt to become a leading producer of nanomaterials and a nanoscience knowledge hub. But whether China is able to leverage off its increasing wealth and funnel this into global leadership in science and technology, in large measure depends on still nascent regulatory systems (Jarvis and Richmond 2011).

In terms of regional ethics concerns, the EU debate is clearly broader than the one taking place in China. While toxicity risk is paramount in both, the EU, having already been 'burned' on the issue of genetically modified (GM) food, has an approach that is more oriented towards public concern. For the EU, 'risk' means not only the possibility of 'catastrophic scenarios of new and unpredictable technologies gone awry' but also of public *perceptions* of such potential catastrophes or of possible social problems (Lin 2006).[2]

EU and Chinese approaches to bioethics and S&T ethics differ in terms of the application of the precautionary principle and the extent of public debate.

How much of that difference relates to apparently diametric cultural values, such as Western individualism and so-called Eastern communitarianism? China is often regarded as the country that most clearly epitomises the so-called Asian values. By Asian, 'Confucian' is usually meant, with a 'Confucian' commitment to 'hard work, thrift, filial, piety, and national pride', having encouraged rapid economic growth in much of Asia (Dupont 1996). Amartya Sen (1997), while suggesting the limitations of such a generalisation, i.e. that it would be 'a mistake both to see Confucianism as the only tradition in Asia' and to see Confucius as simply an authoritarian figure, notes however that a Confucian authoritarian image is often taken as synonymous with 'Asia'.

Rather than the contentious term 'Asian values', however, which would for example, not have too much applicability to India, perhaps 'Orientalism' is a more useful term, in that it implies not so much specific values, as a structure of antithesis. Edward Said's (1977) well-known work on Orientalism suggests a dichotomy between 'us and them' exposed by Western views of the East as inferior. The dichotomy is central to both regions, whose self-identification rests partly on definition through opposition – the East is 'what the West is not' and vice versa. The social and cultural values underpinning bioethics debates in China and the EU might suggest that it is a problem created by different ideas of one's social identity. In the West, identity is driven by human rights discourses and the priority of autonomy, whereas in the East, *tao* (harmony) and *ren* (benevolence) appear more highly valued.

'Otherness' is an integral part of Said's theories of how the clash of civilisations, of East and West, has evolved through negation rather than recognition and acceptance of the other's differences.

China, for the purposes of this book, is posited as potentially, in its communitarianism, antithetical to the EU, a useful 'other'.

1.2   Global Cooperation?

Is global consensus really necessary? New interdependencies in S&T development, along with a globalised and increasingly mobile market for S&T products, require some agreement on how these technologies should be governed. Part Two of this

---

[2] See also Ebbesen (2007) and Moor and Weckert (2004).

book looks at the central question of how we can achieve a consensus shorn of the metaphysical basis on what constitutes the good life, the 'good citizenship', or 'good social identity' of the respective groups that enter into the deliberation (Hongladarom 2004).

Global bodies have achieved some progress in addressing concerns about nanotechnology. Europe is increasingly cooperating and competing with China and India, two major emerging economies that are also keen to develop their S&T sectors. It is arguable that new interdependencies between global actors require innovative global approaches to S&T policy or at least a recognition of different local approaches to a global science.

Of course, this begs the wider question: *Can* there ever be such a thing as a 'global approach'? While universalism has become a much less hegemonic, more pluralistic concept in the last 20 years, relativism has also become less 'irresponsible' (Morin 2009). However, this battle between the universal and the local still presents a challenge for policymakers developing an inclusive global and ethical approach to nanotechnology. One solution to the problem is to increase the means of public participation.

1.3   Increased Public Participation

Participatory technology assessment (pTA) in S&T evaluation is a trend that has grown since the 1980s in the EU. pTA can be defined as a drive towards incorporating social concerns into technology assessment through public dialogue incorporating a greater use of laypersons. This trend has been driven by increased public scepticism of scientific research and development (Gaskell et al. 2002). Reports of public dissatisfaction in China following various food safety scandals over milk and gutter oil confirm that scepticism about (a lack of) regulation is not confined to the EU.

The outcome has been a demand for more informed public debate, in addition to greater public involvement in how new technologies should be regulated and funded as well as how they should be applied to everyday life (Doubleday 2007). Policymakers in the EU anxious not to repeat the mistakes associated with the introduction of biotechnology initiatives like genetically modified (GM) foods – namely, that legislation came too late and with too little public engagement and ethical reflection – began to call for the inclusion of ethical and societal impact assessments of potentially transformative new technologies at an earlier stage in policymaking. With public participation, societal expectations can be unmasked or exposed, and the risk/benefit debate adjusted accordingly.

However, global ethics requires global agents. 'Actors and agendas' is a concern within current nanoethics commentary.[3]

---

[3] See *NanoEthics* 7 (2013), which contains several articles on societal challenges of nanotechnology, in particular, Lotte Krabbenborg's article.

pTA 'nanofora' that have been attempted thus far in the EU and China (the latter being somewhat less concerned with this approach as yet) offer a basis for the evolution of a more developed pTA model, one achieved through linking pTA to the field of discourse ethics. Following the view by Ladikas and Schroeder (2005) that global ethics is reliant on such practical fora as platforms for intercultural dialogue and trust-building,[4] pTA can usefully be informed by the dialogue- and process-oriented theories of a discourse ethicist such as Jürgen Habermas.

Discourse ethics has developed around Habermas's argument that social order depends on our capacity to recognise, through rational discourse, the intersubjective validity of different views. It aims to be a consensual ethics, in which only those norms and actions can claim legitimacy that have been subjected to discourse, and when a noncoercive consensus by all affected has been achieved. Thus Habermas (2004) claims that something can be valid when the foreseeable 'consequences and side-effects of its general observance for the interests and value-orientations of each individual could be jointly accepted by all concerned without coercion'. Communicative or discourse action attempts to coordinate purposive action rationally, so that all might come to a reasoned agreement about the truth of a statement or of the rightness of a norm (Habermas 1998; 2004).

The principles Habermas sees as requisite for dialogue are very similar to those of pTA, particularly those of inclusivity and range of representation. Commenting on Habermas, for example, Benhabib (2003) advocates 'practical discourses' in which all who are affected can be 'participants in the discourse through which the norms are adopted'.

Nanotechnology's issues have already been debated through such an approach, as we can see from nanodialogues conducted in Europe over the last 5–7 years, while the 2006 nanodialogues, for example, held in Zimbabwe sought to inspire bottom-up approaches to nanoinnovation by engaging local community groups to assess the appropriateness of nanotechnologies for community needs (Grimshaw et al. 2006).

pTA and discourse ethics are mutually beneficial for the purpose of this study in terms of Habermas's notion of 'universalisation'. The focus of discourse ethics is what is 'equally good for all', of selecting 'from the mass of evaluative questions those action-related conflicts that can be resolved with reference to a generalizable interest' (Habermas 1992). Habermas asks the basic question of global ethics, of how different views (particularly of social order) can be universally understood and reconciled, perhaps within an 'ideal community' of communication, one that may be global. His 'universalisation principle' offers a useful methodology for examining how a pTA model might work globally.

A model of global engagement is one of global pTA, oriented towards universalisability. It is a model that requires three components:

---

[4] Doris Schroeder and Miltos Ladikas have looked at the issue of globalisation from the rather more practical issue of global basic income, utilising the ideas of Pogge on resource dividends.

- An understanding of the issues that transcends cultural differences
- A pTA structure with clear procedural guidelines and that can affect/effect policy
- And 'virtuous' citizens able and willing to participate in such a process and achieve consensus

Examining the idea of 'global virtues', we can look at Aristotelian, or Western-centric virtues, Habermasian 'virtues' or 'skills', and Chinese *jen* or *ren* ethics, encapsulating such ideas as non-maleficence, respect for others, 'human hearted-ness', benevolence, love, compassion, kindness, and humanity (Chan 1969).

This forms the basis for the concluding section, on whether participants in such a Habermasian-inspired discourse/pTA process might be able to achieve a 'dialogic form' or 'dialogic self', a concept informed to a degree by Korsgaard in her work on agency and self-constitution, Appiah on the dialogically constituted self, Gutmann and Thompson on first- and second-order values, and Charles Taylor (1992) on recognition and difference:

> We give due acknowledgment only to what is universally present – everybody has an iden-tity – through recognising what is peculiar to each. The universal demands an acknowledge-ment of specificity…in the case of the politics of difference, we might say that a universal potential is at its basis, namely, the potential for forming and defining one's own identity.

As Taylor puts it, identity is created dialogically through our interactions with others. Through such intersubjectivity, we develop new vocabularies of comparison, discover shared aspects of identity, and achieve, hopefully, an 'intercultural person-hood' that 'includes a vital component of an outlook on humanity that is not locked in a provincial interest of one's ascribed group membership', as Beck (2002) has stated, and that may lead, perhaps, to a global ethics.

References

Aly W (2007) People like us. How arrogance is dividing Islam and the West. Picador, Sydney

Arnall A (2003) Future technologies, today's choices. Available via Greenpeace Environmental Trust. http://www.greenpeace.org.uk/MultimediaFiles/Live/FullReport/5886.pdf

Ball P (2003) 2003: nanotechnology in the firing line. Available via Nanotechweb. http://nanotech-web.org/cws/article/indepth/18804. Accessed 18 Dec 2014

Beck U (2002) The cosmopolitan society and its enemies. Theory Cult Soc 19(1–2):12–44

Benhabib S (2003) The claims of culture. Equality and diversity in the global era. Princeton University Press, Princeton

Berne RW (2004) Towards the conscientious development of ethical nanotechnology. Sci Eng Ethics 10:627–638

Chan WT (1969) A source book in Chinese philosophy. Princeton University Press, Princeton

Chou K (2007) Biomedtech Island Project and Risk Governance. Paradigm conflicts with a hidden and delayed high-tech society. Soziale Welt 58:123–143

Doubleday R (2007) Risk, public engagement and reflexivity: alternative framings of the public dimensions of nanotechnology. Health Risk Soc 9(2):211–227

Drexler E (1996) Engines of creation: the coming era of nanotechnology. Fourth Estate, London

Dupont A (1996) Is there an 'Asian way'? Survival 38(2):13–33

Ebbesen M (2007) Nanoethics – not from Scratch. Paper presented at the nano ethics workshop, University of Aarhus, Denmark, 22–23 September 2007

ENSAA (2011) Nanotechnology and society: moving upstream. http://www.ensaa.eu/index.php/innovation/106-nanotechnology-moving-upstream.html. Accessed 18 Dec 2014

Gaskell G, Thompson P, Allum N (2002) Worlds apart? Public opinion in Europe and the USA. In: Bauer MW, Gaskell G (eds) Biotechnology – the making of a global controversy. Cambridge University Press, Cambridge, pp 351–375

Gaskell G, Stares S, Allansdottir A, Allum N, Castro P, Esmer Y, Fischler C, Jackson J, Kronberger N, Hampel J, Mejlgaard N, Quintanilha A, Rammer A, Revuelta G, Stoneman P, Torgersen H, Wagner W (2010) Europeans and biotechnology in 2010. Winds of change? A report to the European Commission's Directorate-General for Research. Available via European Commission. http://ec.europa.eu/public_opinion/archives/ebs/ebs_341_winds_en.pdf. Accessed 18 Dec 2014

Grimshaw DJ, Stilgoe J, Gudza LD (2006) The role of new technologies in potable water provision: a stakeholder workshop approach. Can nanotechnologies help achieve the millennium development target of halving the number of people without access to clean water by 2015? http://practicalaction.org/docs/ia4/nano-dialogues-2006-report.pdf. Accessed 18 Dec 2014

Habermas J (1992) Discourse ethics, law and Sittlichkeit. In: Dews P (ed) Autonomy and solidarity: interviews with Jürgen Habermas. Verso, London, pp 245–271

Habermas J (1998) The inclusion of the other. Studies in political theory. MIT Press, Cambridge

Habermas J (2004) The theory of communicative action. Reason and the rationalization of society, vol 1. Polity Press, Cambridge, pp 285–287

Hongladarom S (2004) Asian bioethics revisited: what is it? and is there such a thing? Eubios J Asian Int Bioeth 14(6):194–197

Illuminator I, Miller G (2010) Nanotechnology, climate, and energy: over-heated promises and hot air. Available via Friends of the Earth. http://www.foe.co.uk/sites/default/files/downloads/nanotechnology_climate.pdf. Accessed 18 Dec 2014

Jarvis D, Richmond N (2011) Regulation and governance of nanotechnology in China: regulatory challenges and effectiveness. Available via Eur J Law Technol 2(3). http://ejlt.org//article/view/94. Accessed 18 Dec 2014

Joy J (2000) Why the future doesn't need us. Available via Wired. www.wired.com/wired/archive/8.04/joy_pr.html. Accessed 18 Dec 2014

Krabbenborg L (2013) DuPont and environmental defense fund co-constructing a risk framework for nanoscale materials: an occasion to reflect on interaction processes in a joint inquiry. NanoEthics 7:45–54

Ladikas M, Schroeder D (2005) Too early for global ethics? Camb Q Health Care Ethics 14(4):404–410

Lin A (2006) Size matters: regulating nanotechnology. Available via Social Science Research Network. http://ssrn.com/abstract=934635. Accessed 18 Dec 2014

Moor J, Weckert J (2004) Nanoethics. In: Baird D, Nordmann A, Schummer J (eds) Discovering the nanoscale. Ios Press, Amsterdam, pp 301–310

Morin M (2009) Cohabiting in the globalised world: Peter Sloterdijk's global foams and Bruno Latour's cosmopolitics. Soc Space 27:58–72

Nanotechnology Now (2007) Nanotechnology in China is focusing on innovations and new products. Strong growth. http://www.nanotech-now.com/news.cgi?story_id=22584. Accessed 18 Dec 2014

Said E (1977) Orientalism. Available via One Democratic State Group (ODSG). http://www.odsg.org/Said:Edward(1977)_Orientalism.pdf. Accessed 18 Dec 2014

Schummer J (2007) The impact of nanotechnologies on developing countries. In: Allhoff F, Lin P, Moor J, Weckert J (eds) Nanoethics: the ethical and social implications of nanotechnology. Wiley, Hoboken, pp 291–307

Sen A (1997) Human rights and Asian values. What Kee Kuan Yew and Lee Peng don't understand about Asia. The New Republic. http://www.hmb.utoronto.ca/HMB303H/weekly_supp/week-02/Sen_Asian_Values.pdf. Accessed 18 Dec 2014

Stephenson N (1995) The diamond age. Viking, London

Taylor C (1992) Multiculturalism and 'The politics of recognition'. Princeton University Press, Princeton

Chapter 2
What Is Nanotechnology, and What Should We Be Worried About?

> *Nanotechnology quests, like all newly developing technological ambitions, are also quests for personal fulfilment. (Berne 2004)*
>
> *How can such a brave new science, one that is so full of potential that it has been called the 'Next Industrial Revolution' by governments and scientists, not also impact our relationships, society, environment, economy or even global politics in profound ways. (Lin et al. 2007)*

Abstract This chapter compares issues identified in the nanoethics field in China and the EU, such as enhancement, health, environment, military usage, and privacy concerns, as well as the significance of public perception of nanotechnology. The 'speculative' approach, in terms of which science fiction novels' depiction of nanotechnology issues is used as a medium for debate on nanoethics, occupies the second half of this chapter.

Keywords Nanoethics • China • EU • Enhancement • Science fiction • NELSI

For members of the public, nanotechnology (or, as it is also labelled, nanoscience, nanoscale research, or nanotechnoscience) is a term only vaguely understood and is often confused with the popular and inaccurate trend of labelling anything in technology with very small features as 'nano'. Defined properly, nanotechnology is the 'engineering of functional systems at the molecular level' (Centre for Responsible Nanotechnology 2002), with one nanometre (nm) being one billionth of a metre.[1] The European Commission suggests 'nanosciences and nanotechnologies' to distinguish between the science and its technological application; the label 'nanotechnology' seems the most relevant (as well as the most broadly used) term for

---

[1] To put this in context, a human hair is 80,000 nm wide. Although there has been some debate about the precise size range within which a structure may be called nano, this is not of particular relevance to this study. The relevant range is usually 1–100 nm, but can extend to below 0.1 nm and above 100 nm; the upper limit of 100 nm for nanotechnology has however been called arbitrary, and Allhoff (2007) argues, such a limit ignores the dimensionality that nanowires and nanofilms demonstrate.

© Springer International Publishing Switzerland 2015

S. Dalton-Brown, *Nanotechnology and Ethical Governance in the European Union and China*, DOI 10.1007/978-3-319-18233-9_2

discussing nanomanufactured technologies that affect our lives (European Commission 2004).

From artificial coatings made by nanothin layers that, when used on steel, could have delayed the 9/11 collapse by over 20 min, to rust nanoparticles to which arsenic oxide would bond, thus purifying water, to nanoparticles used in household refrigerators, it is evidently a technology with the potential to affect most major industries.

In a 2003 Greenpeace report, Arnall, noting the breadth of the field, suggests that nanotechnology is a catalyst for general scientific advancement. For example, if large-scale carbon nanotube production does become a reality, it will be equivalent 'to the development that kick-started the industrial revolution', for nanotube-based material can offer a substance 50–1,200 times the strength of steel at one sixth of the weight (Arnall 2003). Thus

> ...complex nanotech products will change five very different parts of the economy: the chemical industry, textiles, the construction industry, the food/agriculture industry, and robotics. The chemical industry may be transformed into a nanomaterials industry. Nanotech may change the competitive factors in the textile industry, create new products for the building industry along with making the food and agriculture industry more efficient. (Gasman 2006)

Thus possible applications of nanotechnology range across:

- Better construction and manufacturing methods
- Improved water filtration products
- Better energy storage and conversion
- Improved IT memory and storage
- Better paint and textiles that are resistant to stains, UV, and bacteria
- Improved military/security technology (battle suits, surveillance devices)
- Improved food packaging for storage
- More effective cancer treatments
- Improved pesticides and fertilisers

Nanotechnology is often examined not only in terms of its potential and practical applications, but also for its transformative potential. Under the heading of nanotechnology we tend to think of engineering at the scale of the very small; however, we may also remember utopian (or dystopian) visions, such as that in Eric Drexler's 1986 *Engines of Creation*, of tiny assemblers capable of creating any desired macroscopic object. With such technology we could, theoretically, create *anything*, a concept explored by Neal Stephenson in his novel *The Diamond Age* (1995), which depicted a world of ubiquitous nanotech where molecular assemblers can create basic goods, free to anyone who requests them, thus eradicating starvation and homelessness (Aiyu Niu 2008).

It is a vision of 'nanotech synthesizers in factories, then restaurants, and finally in homes' of foods that 'could simply be synthesized cooked, with no need for a stove', while the dishes are also 'synthesized along with the food, and then simply dropped dirty into the recycler' (Storrs Hall 2005). This kind of future could enable consumers to enjoy not only plentiful, but 'interactive' food that would allow for modification based on their own nutritional needs or tastes: 'thousands of nano-capsules containing flavour or colour enhancers, or added nutritional elements (such as vitamins), would remain dormant in the food and only be released when triggered by the consumer' (Dunn 2004).

Even more radically, the UK science fiction series *Red Dwarf* episode 'Nanarchy' (March 7, 1997) shows nanobots recreating body parts, in this case rebuilding a character's missing arm. Such visions stimulate much 'nano hyperbole' (Ellen Mitchell 2007), claims that nanotechnology will provide an answer to the problem of how to continue to satisfy all of our pressing material needs. One response to this might be to distinguish between futuristic and near term visions of nanotechnology. Certainly, nanobots and cell reprogramming are not as yet anywhere near realisation, and molecular assembly is not expected for decades (if at all) (Lopez 2004; Davies and Macnaghten 2010; Dupuy 2010). Thus, we might distinguish between evolutionary or 'incremental' technology and disruptive or radically transformative technology (Wilsdon distinguishes between 'shallow' and 'deep' technologies) (Wilsdon 2004).

The radical novelty of nanotechnology becomes apparent when we see the emergence of 'unusual physical, chemical and biological properties' and of new materials, such as the 'buckyball', discovered in 1985. This new material has many potential applications in medicine, photo-optics, and energy; for example, bucky-balls can be used to store hydrogen at high densities, allowing hydrogen to become economically competitive with gasoline as car fuel, thus solving one of the challenges faced by the alternative energy industry (Mick 2008).

The discovery of buckyballs was followed by that of carbon nanotubes in 1991. The latter, as they generate no heat during conduction, have the potential to transform thermal management products (i.e. more efficient cooling), while their considerable tensile strength has already allowed them to be used as strengthening agents in polymers and concrete – they might eventually provide lightweight material to build spacecraft or even allow the building of a 'space elevator' to move people and materials out of earth's orbit. In medicine, nano-tubes could increase the rate of healing by creating a 'scaffold' to assist bone growth, act as cellular 'needles' to assist targeted drug delivery, or be placed in sensors to detect pollutants.

In the energy field, Australia's national science agency, the CSIRO (Commonwealth Scientific and Industrial Research Organisation), in conjunction with two Australian universities, is developing nanochemical sensors to enhance discovery rates of untapped oil or gas deposits beneath the seabed, in a bid to prepare for predicted oil and gas shortages in the future. Petrochemical companies suggest that nanotechnology will enable far greater rates of extraction from existing reserves (through creating lighter yet stronger 'proppants' that are forced into drilling fractures) (Johnson 2010).

The potential applications seem endless.

Yet much of what is sold as nanotechnology is just a redefinition of existing science. Consider sunscreens, which use miniature colloids, an application of general chemistry (Moor and Weckert 2004). Or computer chips assembled in familiar ways, though now merely on a microscale – is this really a new technology (Preston et al. 2010; Rodgers 2006)?

The argument as to whether nanoscience is in fact a *new* science can be extended into a discussion of nanomedicine, arguably a 'continuation of the understanding of physiological, biochemical, cellular, genetic, and disease processes' (Hook 2007). The term 'nano-enabled' might in fact be more accurate than 'nanotechnology' in some cases.

Nanoscience can also be defined as a convergence technology, an interdisciplinary science, or amalgamation of established disciplines (Preston et al 2010). Biomolecular nanotechnology has research potential from molecular biology and biophysics to applications in biosensing, biocontrol, bioinformatics, genomics, medicine, computing, and information storage and energy conversion.

A 2006 study of nanotechnology patents suggests the USA leads in four out of the top five top fields: nanotubes, plasmonics (optical data transmission), self-assembly, and (anti)reflection. Europe was also shown to be catching up in plasmonics. Nanotech scanning probes enabling the production of better microchips and other microelectronics are likely to remain the largest focus for patents (Schmid et al. 2006).

One of the biggest areas of research is nanomedicine. Although the ultimate goal, *Red Dwarf* style of monitoring, repairing, and improving all human biological systems, is far off, nanomedicine is predicted to lead to both major health improvements and very good economic returns in the near to medium future. Adams and Williams (2006) note that the US National Institutes of Health (NIH) Roadmap's Nanomedicine Initiative lists its major goals as:

> ...finding ways to 1) search out and destroy early cancer cells, 2) remove and replace broken cell parts with nanoscale devices, and 3) develop and implant molecular pumps to deliver medicines.

Nanoscale science is potentially useful for diagnostics, tailor-made drug design, and cell targeting. A possible application is that of gold nanoshells being targeted onto a tumour. When a laser is directed onto the tumour, it will heat and destroy the gold while leaving the healthy tissues (with no gold nanoparticles attached) cool and unaffected – a huge benefit compared to chemotherapy.

A further application will be the development of 'lab-on-a-chip technology' which, at the nanoscale, allows for biological samples to be tested for hundreds of known conditions. Such field labs can allow for far greater access to medical help and to swifter diagnosis and better disease treatment, particularly in developing countries.

Another example is that of pharmaceutical companies using cost-saving nanotechnology to reduce the size of drugs due to increased surface-to-volume ratios and reactivity. The benefits have been summarised as follows:

> Nanotechnology can enhance the drug discovery process... It will also result in reducing the cost of drug delivery, design and development and will result in the faster introduction of new cost-effective products to the market. (Bawa and Johnson 2008)

Nanodrugs may be more efficiently administered transdermally rather than orally. Better time-release drug products can be manufactured for more calculated long-term dosages (Allhoff et al. 2010).

More than 30 countries had national nanotechnology activities in 2001, while others had some evidence of individual or group research, thus being considered 'developing or transitional'. These were countries ranging from Argentina, Belarus, and Bulgaria to Turkey, Uruguay, and Vietnam (Maclurcan 2005).

The ETC Group (Action Group on Erosion, Technology, and Concentration) has estimated that 60 countries have state-level 'nanotech initiatives, including new-comers Nepal, Sri Lanka and Pakistan' – a doubling of activity in 10 years (ETC Group 2010). A more recent comparison of countries according to whether their research activity is globally dominant in commercial terms shows Japan, Germany, South Korea, and Taiwan as leading the way, with China and the UK also in the top six of those dominating the market with nanoproducts (Hwang 2010).

2.1 Products for the Public

Research activities indicate that nanotechnology may eventually realise many of science fiction's predictions. Yet as far as the public is concerned, the current or imminent *results* of the above research, namely, the products, are what count. The Project on Emerging Nanotechnologies estimated in a 2008 report that over 800 manufacturer-identified nanotech products were available, with new ones coming onto the market at a rate of three to four per week.[2] According to a 2009 report, there were 600 products in the market that producers claim are based on nanotechnology (Decker and Li 2009). It was estimated that US$32 billion worth of products incorporating nanotech were sold globally in 2006, and by 2014, 2.6 trillion US dollars in manufactured products will be 'nanoproducts', making up 15 % of total global manufacturing (Kimbrell 2007). This figure is, however, debatable – a 2007 analysis described these large figures as a purely sensationalist attempt to hype an artificially constructed market (Berger 2007).

The list of current products contains a considerable range of items, from cosmetics and sunscreens to clothing with dirt-repellent nanofinishes, to solar technology, to scratch-proof paint. In Israel, sol-gel nanotechnology is being used to make a more efficient acne cream. Nanocosmetics is already a large market, with L'Oreal owning more than 200 patents allegedly utilising the ability of nanoparticles to penetrate deeply into the skin. Sol-gel technology is also used in the production of perfume and of more efficient surface coatings with improved corrosion and abrasion protection (Pagliaro 2010).

Nanopaints with improved insulating properties can reduce a household's energy bill, while the heat conductivity of solar panels may be improved with nanoen-

---

[2] The Project on Emerging Nanotechnologies has an online inventory of products at http://www.nanotechproject.org/cpi/.

hanced materials (Li and Somorjai 2010). Titanium dioxide, a brilliant white pigment used in paints, is transparent at the nanoscale and has a photocatalytic effect that helps sunlight break down dirt. When applied to glass, the result is 'self-cleaning' windows (Twist 2004). Then there is kitchenware, stain-resistant clothing, insulated footwear (Wolfe 2005a, b), bone tissue implants (to supersede traditional orthopaedics), and one of the earliest products Wilson Double Core tennis balls, which keep their bounce longer (being coated internally with a nanocomposite that reduces air leakage).

Take Thailand as one example of a varied nanomarket: the products of focus for manufacturers are waterproof and more durable silks, 'smart packaging' to monitor and maintain the state of food, more productive wine fermentation, 'self-sterilising' rubber gloves, and new car body materials (Changsorn 2004).

On the topic of smart packaging for food, nanotechnology has the potential both to cut waste due to improved packaging[3] and to reduce bacteria that cause food poisoning (Science Daily 2008). Food packaging is of importance for countries that may not have efficient supply chain structures, in terms of reducing food wastage during transportation. In the food industry alone, experts estimate that nanotechnology will be incorporated into $20 billion worth of consumer products by 2010 (Helmut Kaiser Consultancy 2006). Companies such as H.J. Heinz, Nestlé, Hershey, Unilever, and Kraft are investing heavily in nanotechnology applications (Wolfe 2005a, b).

Nanotechnology could have a huge impact on developing countries, having the potential to advance agricultural productivity through the genetic improvement of plants and animals, delivery of genes and drug molecules to specific sites at cellular levels in plants and animals, and gene technologies for plants and animals under stress conditions (Kuzma and VerHage 2006). Iron nanoparticles that apparently break down harmful substances in soil could prove environmentally valuable (Schwarz 2009) and so would water filtration devices (extremely fine nanofilters) allowing rural populations in Africa to access clean water at minimal cost.

The 1.6 billion dollars of public money pledged for research in the USA in 2010 indicates the central role the US government thinks nanotechnology will play across several areas of technology. Yet there is also the rather cynical view that comes after spending any time online looking for nanoproducts. 'Nano-tea', allegedly made by extra-efficient pulverising of the tea leaf to so-called 'nanograde' fineness, is just one of these specious products. Some products are also misleadingly labelled, such as the first so-called nanoproduct to be recalled from the market, the bathroom cleaner 'Magic Nano', later found in fact to contain no nanoparticles (Pagliaro 2010). More seriously, it has been suggested that nanotechnology 'is going underground' as a result of the controversy surrounding nanotechnology, leading manufacturers to remove any mention of nanomaterials from their products while continuing to use nanotechnology processes (Nanowerk 2009).

---

[3] Bayer Polymers has developed a packaging film enriched with silicate nanoparticles that 'massively reduce the entrance of oxygen and other gases and the exit of moisture, thus preventing food from spoiling' Joseph and Morrison (2006).

With this broad range of applications, it is clear that the public will inevitably have to engage (indeed, already is engaging) with nanotechnology. Let us now look at some of the ethical issues stimulated by the emergence of such a potentially transformative technology.

2.2 Approaching Nanoethics

The role of nanoethics is 'to engage in conversation with others about what kind of world should be constituted'. (Johnson 2007)

When approaching the issue of the effects of a new technology, one standard approach is that of 'ELSI', which examines the 'ethical, legal and societal implications' of new technology. This acronym appeared in 1989 in the West as part of the Human Genome Project, the basic goal of such an approach being to forestall adverse effects associated with biotechnology, to sponsor research and conferences, and also to make policy recommendations. The temptation for commentators on nanotechnology's implications for society has been to add an 'N' at the start – hence a 'NELSI' approach (Fisher 2005).

This makes it sound a little easier than it is, as one issue with nanotechnology is not merely its extensive reach of applications, or the idea of 'adverse effects', but its 'unknowable' quality – something that commentators have even labelled 'noumenal' (more intellectually than empirically understood). Nanotechnology may also require a paradigm shift in the *way* we think about it; such a new science may require a radically new approach, for '*noumenal* technology recedes into the uncanny otherness of nature and resists our attempts to make it an object of experience and knowledge' (Nordmann 2005). Thus Dupuy and Grinbaum (2006) argue for insufficiency in all of the three main, traditional theoretical approaches (of consequentialism, virtue ethics, and deontology):

Virtue ethics is manifestly insufficient since the problems ahead have very little to do with the fact that scientists or engineers are beyond moral reproach or not. Deontological doctrines do not fare much better since they evaluate the rightness of an action in terms of its conformity to a norm or a rule… As for consequentialism – i.e. the set of doctrines that evaluate an action based on its consequences for all agents concerned – it treats uncertainty as does the theory of expected utility, namely by ascribing probabilities to uncertain outcomes.

However, this statement seems a little naïve, in that ethical theories are rarely used uncritically.

Can one have an ethical debate about technologies that do not yet fully exist (Gorman et al. 2007)? Does such a new science require a new approach and a new set of ethical questions (Ferrari 2010)? Nordmann (2006) argues we cannot predict the future of nanotechnology – all we can say is that it is a 'horizon of expectation in which something unheard of or unspeakable will appear'. Yet Weckert (2007) states sensibly that worrying about predictive assessment is unhelpful, that

> there must be ethical discussion before, during, and after development of the technology. Ethics must be proactive, reactive, and it must occur *pari passu* with the development.

Bostrom (2007) suggests a need to 'develop better high-order epistemic principles for the conduct of scientific research', concluding that 'until we achieve a dramatic enlightenment in our capacity for pragmatic synthesis…we will continue to stake out our ethics and policy paths in the dark'.

Allhoff (2007) has argued that nanotechnology issues are similar to those raised by other technologies, but that should not detract from the need to apply specific ethical attention to nanotechnology – an approach that seems sensible.

Some commentators suggest a compromise – not an entirely new paradigm, but less structured, more laterally innovative, and interdisciplinary approaches to the challenge.

Berne, for example, advocates the use of 'creative moral imagination' (Berne 2005, 2008). Cameron sees virtue in nanoethics detaching itself from bioethics to adopt a deliberately broader approach (de S. Cameron 2007). Gordijn (2005) argues that nanotechnology's breadth may require some 'new questions', while Swiestra and Rip (2009) suggest that new technology may lead to the 'co-evolution' of ethics and new technologies.

Whereas many commentators take a consequentialist approach, Lewenstein (2005), for example, argues that instead of looking at areas of nanotechnological *impact*, one should look at the four broad questions of fairness, equity, justice, and power across nanotechnology areas. His argument is developed by Bennett-Woods (2008), who also looks at issues of competing loyalties, such as security versus privacy or profit versus equity.

In practical terms, the nanoethics debate could be reduced to the following set of challenges:

- There may be a need for anticipatory governance.
- Informing the public of nanoissues becomes a particularly complex task. Doubleday, reviewing six public engagement projects on nanotechnology in the UK and USA in 2005–2006, argues that the nanotechnology 'framing of public engagement is both too broad and too narrow to allow for a fully articulated public discussion' (Doubleday 2007). The questions of nanotechnology may not as yet have been fully articulated
- And for 'nanoethicists', the challenge is to identify and debate complex nanoissues – possibly by developing more 'creative' methodologies than those already being utilised.

In response to the last point, we might begin by looking to science fiction as a means of isolating nanoethical issues.

2.2.1 Hypothetical Nanoethics? Science Fiction and Nanotechnology

Mind uploading would be a fine thing, but I'm not convinced what you'd get at the end of it would be even remotely human. (Stross 2012)

We begin our discussion of nanoethics and our eventual discussion of how Eastern and Western approaches to the issue reflect differing ethics with a quick trip through nano-themes as handled by science fiction writers in the West. It should be noted that (as discussed in the *Introduction*) for the purposes of this study, by 'West' I refer chiefly to the EU but also at times briefly to the USA and by 'East' principally to China.

Richard Feynman (1960) has suggested that the scientific and visionary aspects of nanotechnology have been interconnected from the beginning.[4] Science fiction can in some of its incarnations, such as satiric texts, for example, be an effective forum for ethical debate. Science fiction texts offer the reader contemporary issues in new contexts, thus casting them in fresh light (Milner 2006).

Milburn (2008) suggests in fact that nanotechnology has *depended* throughout its history on a symbiotic relationship with science fiction. Dupuy argues that 'dreams of reason', which 'can take the form of science fiction', 'have a causal effect on the world and transform the human condition' (Dupuy 2007). This is rephrased by Grunwald (2010) as 'explorative nanophilosophy' that has 'something valuable to contribute to future nano-debates, and future applied ethics'. Engineering students have even been set science fiction texts in class to ensure that they understand the 'ideological landscape' of nanotechnology (Toumey 2008).

Of course, the opposite view can be argued. Detractors suggest that this speculative focus can distract from current realities (Nordmann 2007),[5] that 'most nanoethics is too futuristic…at the expense of ongoing incremental developments that are more ethically significant' (Nordmann and Rip 2009). Gordijn (2005) argues that utopian or apocalyptic visions impede nanoethics in their encouragement of far-fetched scenarios, and Keiper (2007) even suggests that what happens in such speculative debates is a short-circuiting of logic, whereby the unlikely becomes inevitable.

Despite this criticism, science fiction's exploration of more speculative aspects of nanotechnology might at the very least afford a useful insight into the public's perception of this complex technology. In an analysis of UK media coverage of nanotechnology over a 15-month period (2003–2004), the most prevalent reference was to Michael Crichton's novel *Prey,* which features deadly nanoswarms (Anderson et al. 2009). Although Crichton's critical gaze appears chiefly aimed at profiteering corporations, the image of a woman controlled by a devouring technology that

---

[4] He was referring principally to the utopian visions of science fiction writer Frank Heinlein.

[5] See also Roache (2008). Roache states that ethicists should 'focus on maximising what is most valuable', regardless of whether it is a current concern or a 'desirable vision of the future'.

exhibits predatory swarm behaviour is a sobering image of the destructive consequences of rushing headlong into a new technology.

Nanofiction can usefully depict those pressing ethical issues that require consideration, delineating areas of concern and offering hypothetical scenarios that highlight contemporary problems. This can help in encouraging or directing public debate. Such issues become quickly apparent from any brief analysis of the last two decades of science fiction, nanotechnology having become a theme in the 1990s in Western literature after its early minimal appearance in such texts as Arthur C. Clarke's 1956 story *The Next Tenants*, describing tiny machines that operate on a microscale, or Robert Silverberg's 1969 short story *How It Was when the Past Went Away*, which shows a form of nanotechnology being used in the construction of loudspeakers.

One should first note however that such an analysis is of limited use for the purposes of this study and its comparison of Chinese/EU attitudes to nanotechnology. This is due to the fact that *Eastern* science fiction is a rather different genre. If one compares Western science fiction novels from the USA and the UK with those published in China, it immediately becomes apparent that the latter are not necessarily a useful speculative forum. As in Russia, China has struggled with a genre that simultaneously promotes technological superiority and innovation while also showing a deep level of satire. There is a certain nervousness in China even when it comes to time travel TV shows, criticised for promoting 'feudalism, superstition, fatalism and reincarnation' (Sebag-Montefiore 2012). Two recent science fiction novels from China, Han Song's *2066: Red Star over America* (2000) and Chan Koonchung's *The Fat Years* (2009), thus only cautiously examine China's superpower status. Liu Cixin perhaps conceals his message within a depiction of aliens' attempts to control the minds of the Chinese populace in his trilogy *Three Bodies* (2007–2011).

The East does have a tradition of translating classic science fiction novels from English – Jules Verne being one popular choice. Yet Verne's popularity in the East is probably due to the way his novelistic style adapts well to the 'industrial' focus of Chinese science fiction, i.e. that celebrates technology rather than tackling ethical problems (Chen 2007).

The nanoissues raised by Western science fiction writers can be listed as follows:

(a) Nano-based consumerism is not necessarily socially beneficial – the 'lotus-eater' scenario.

(b) Technology might outstrip our ability to control it – the doomsday/dystopia scenario. Examples include Crichton, Stel Pavlou, and Robert Ludlum, who have used nanotechnology (*Prey, Decipher, The Lazarus Vendetta*) as a useful part of the thriller's nemesis theme, in terms of which 'the enemy' always has dark forces, whether merely violent, or scientifically terrifying, at his command.

(c) Nanoenhancement might lead to new types of 'super' human beings, creating increased social elitism or nanosoldiers. It might even destroy our humanity as illustrated by the 'superhuman', 'cyborg', and 'Frankenstein' scenarios.

(a) *Lotus-eating in the nanofuture*

It is conceivable that the best-known novel on nanotechnology, apart from Crichton's *Prey,* is Neil Stephenson's *The Diamond Age* (1995), already mentioned as depicting a world of ubiquitous nanotech where molecular assemblers can create basic goods that are freely distributed to anyone that requests them. Yet Stephenson's world of abundant nanotech is socially stratified and stagnant. In such a world of plenty, there is little to strive for – apart from the freedom denied to the poor, kept docile through nanomanufactured goods and denied any say in their lives. Similar to Stephenson's vision of autocratic Victorians, Charles Stross' *Singularity Sky* (2003) depicts the New Republic's iron grip on its populace as enabled through having control of so-called cornucopia machines, which are a type of nanotech assembler factory. For both of these novels, nanotechnology in the form of molecular assembly seems to be the modern equivalent of the Roman Empire's bread and circuses, intended to keep a population well fed and docile.

Interestingly for the purposes of this book (although digressive in this current discussion), Stephenson (1995) introduces occasional references to the clash of Western and Eastern cultures. A character remarks at one point how 'poisonous' Western technology has been to China:

> Just as our ancestors could not open our ports to the West without accepting the poison of opium, we could not open our lives to Western technology without taking in Western ideas, which have been as a plague (sic) on our society. The result has been centuries of chaos. (Stephenson 1995)

Nancy Kress' 2008 story *Nano Comes to Clifford Falls* develops the idea of nanotechnology having reduced the human drive to achieve. Her utopia of abundance at first looks rather appealing; however, this quickly changes into a dystopia. The molecular assemblers given to the town of Clifford Falls can make the protagonist's neighbour a Porsche and foodnano makes everyone's meals, until someone hacks into the nanomachine and all it will make is garbage cans, the mayor has to outlaw the making of nanoliquor, and eventually all the teachers quit and home schooling has to be reintroduced. Crime increases due to the lack of police officers and eventually some towns fall into anarchy.[6] Thus, nanodestruction is brought about by what at first seems like a good idea, namely, the reduction of poverty through nanomanufacturing.

(b) *Unleashing nanoterror and/or creating the 'nano-panopticon'?*

The nano-doomsday scenario will be familiar to readers of pulp fiction, or watchers of B-movies in which cackling scientists maniacally work on destructive devices.

---

[6] Go to http://escapepod.org/2006/10/12/ep075-nano-comes-to-clifford-falls/ for a podcasted reading of the story.

Although there are some positive texts, such as Damien Darby's *Nano Saviour* (2012), in which nanomachines create an artificial intelligence that restores ecological stability, science fiction writers opt for more destructive and dystopian nanofutures – probably because impending disaster makes for a more thrilling plot. Drexler's famous end-of-the-world scenario is fairly popular: arguing that molecular assembly should be treated with caution, as assembler-based life forms, being more efficient than natural organisms, could destroy the biosphere. He even suggests that runaway nanotechnology could turn the universe into 'grey goo' (although in fact he uses the image of dust to signify the state of matter once the biosphere has been consumed, a state that is indeed probably grey, but not particularly gooey). Other writers have replicated this scenario in various doom-laden ways. Kurt Vonnegut's 1962 *Cat's Cradle* hypothesises a form of nano-permutation of water called 'ice-nine', which is solid at room temperature and which inevitably gets released into the environment, bringing about the end of the world (Koepsell 2010). And in Walter Jon Williams' *Aristoi* (1992), nano-apocalypse has already occurred – the universe is on its 'backup version', post the incident of 'Mataglap Nano', a grey goo-style disaster which originated in Indonesia.

Vonnegut's apocalypse is created unintentionally, but what of more intentionally destructive uses of nanotechnology? The US air force noted the following important areas of nanoresearch in 2003: 'increased information capabilities; miniaturization of systems; new materials resulting from new science at these scales, and increased functionality and autonomy'. Taken together, this suggests the creation of autonomous war machines (National Research Council 2003). The latter sounds potentially rather doom-laden, reminiscent (at a stretch) of Kevin J. Anderson's 1993 novel *Assemblers of Infinity* that shows earth ironically making plans for war against 'voracious' nanos (the familiar runaway technology scenario).

In David Nelson's 2011 thriller *Nano War*, the protagonist works with Al-Qaeda on nanotechnology-based weapons of mass destruction that could soon be unleashed against America. The 2007 test of a Russian thermobaric bomb allegedly revealed the development of a more efficient explosive manufactured through nanotechnology and might raise the dark image of suicide bombers with even more destructive power strapped to their bodies (BBC News 2007). However, Russia's growing nanotech focus appears to be far broader than just military use, with eight priority directions of nanotechnology development defined as necessary for the Russian economy.[7] There might be benefits as well, such as in more efficient battlefield wound treatments made available through nanoparticles.

---

[7] This can be argued from the growth in nanotechnology papers of 11.88 % between 2000 and 2007 in Russia. However, compare this with 33.51 % in India over the same time period. See Liu et al. (2009). Russia stated in 2008 that it would invest up to 286 billion rubles ($9.4 billion US, approximately, at that time) in nanotechnology projects across the period 2008–2015; see http://en.rian.ru/russia/20090918/156174098.html and http://english.pravda.ru/science/tech/28-04-2010/113140-nanotechnology-0. Russia's centralised agency, 'Rusnano' (created in 2007), has a commercial focus on investment projects in the aerospace, nuclear, and energy areas, running projects in the fields of solar energy and energy, nanostructured material, medicine and biotech, mechanical engineering, and opto- and nano-electronics.

Many science fiction texts use nanotechnology as a plot device while offering cautionary tales of scientific hubris. Others are more insightful in terms of their vision of the societal consequences of a new technology. Chris Howard's *Nanowhere* (2010) depicts a utopian world of 'socialised medicine based on the coming miracle of molecular engineering'. Yet ultimately this leads to dictatorship, political prisoners, and sterilised women.

One less frequent aspect of the nanofuture in science fiction is that of 'nano-panopticism' (Mehta 2002). This occurs as a result of nanotechnology being used to create ultrasmall surveillance devices. In fact, a nanodevice capable of powering itself by harvesting energy from vibrations while at the same time wirelessly transmitting data over long distances was announced in 2011. The team behind it suggested its positive outcomes for medical technology and the environment:

> The idea that something so small might be able to transmit data across distances could lead to new generations of medical sensors powered by a person's own blood flow, environmental sensors powered by the ebb and flow of atmospheric air, and wearable sensors that run and transmit on the power leftover by the wearer's own footsteps. (Dillow 2011)

Less optimistically, devices of this type might have additional implications for privacy and civil liberties, for 'nanotechnologies will enable more sophisticated monitoring and surveillance technology' (Weckert 2009). Yet science fiction writers, possibly because the 'Big Brother' idea of increased monitoring is such a staple of the genre, do not seem to have engaged with the question of whether the advent of nanotechnology has made this situation markedly more challenging in ethical terms (Nature Nanotechnology 2007). This is odd when we look at a 2004 survey in the USA that reported how 'losing personal privacy' was the most salient concern about nanotechnology. A survey taken a year later detected a similar public opinion (Cobb and Macoubrie 2004; Scheufele and Lewenstein 2005). But one can say that the privacy debate has been ongoing for some time, in various forms. For example, in Japan there has been consistent interest shown in the issue of medical records' confidentiality, whereas in the USA, privacy debates have become subsumed under the problem of terrorism and security (Stanton 2006). Consequentially, nanoethicists such as Bennett-Woods have asked whether personal privacy constitutes a basic human right, or if it should be subject to the 'greater good' of national safety and security. If this is the case, then 'what constitutes a morally unacceptable threshold of threat or risk to justify loss of privacy' (Bennett-Woods 2008)?

In cyberpunk novels – a branch of science fiction depicting societies dominated by technology – surveillance is taken for granted within the fictional setting. These worlds contain a ubiquitous level of nanotechnology, accelerating the datastream in which the characters now live. They are worlds of fast information manipulation.

The ramifications of nanosurveillance are broad, and there are various ethical dangers that might be created. It would be possible to increase economic targeting of consumers due to monitoring devices. Toumey (2007) predicts improved fingerprint recognition that could allow one to derive 'lifestyle intelligence' based on a person's fingerprints, such as whether he/she is a smoker.[8] Perhaps more worrying

---

[8] See also Sametband et al. (2005).

might be so-called genetic discrimination based on increased nanodiagnostic information being available to, say, insurance companies. Implanted biochips, although intended to monitor tumours or activate targeted therapy, may reveal further information that is detrimental to the patient's insurance status. Nanotechnology may offer the potential for faster and thus more cost-effective genome mapping; thus, population screening of neonates might become freely available. This raises ethical issues already identified and relating to civil liberties including possible restrictions on insurance, employment, and even reproduction. The 1997 film *Gattaca*, which featured fast DNA sequencing based on a nanotechnology device and which allowed for both mentoring of and discrimination against so-called 'imperfect' humans, provides one extreme vision of a society based on nano-enabled elitism. And in his *Wondergenes*, Mehlman suggests another elitism, offering the scenario of a Harvard Business School Class made up of applicants selected according to their genetic profiles, suggesting the creation of a new 'genobility' (Mehlman 2003).

This leads us into the topic of one particular form of potentially runway or destructive nanotechnology – human enhancement.

(c) *Nanosoldiers, nanoelites, cyborgs, and Frankensteins*

Doomsday scenarios aside, the major area of interest to science fiction writers appears to be that of human enhancement – perhaps obviously, given that writers of science futures are often interested in the next evolutionary step for humankind. What *kind* of enhancement, however? Moral? Physical? Cognitive? Savulescu (2009) suggests in writing on genetic enhancement that we might even improve our *moral* behaviour through enhancement. Nanomedicine is presented as the 'key that will unlock the indefinite extension of human health and the expansion of human capabilities' (Freitas 1998). Does one not have a moral imperative to become the 'best one can'? Surely society as a whole benefits, the more its citizens improve? Or is this 'playing God' with potentially disastrous socioeconomic consequences? (Belt 2009) The vision of molecular assembly, that of life 'visualized as a do-it-yourself kit', that 'implies that we can also take the world apart and rebuild it to our own taste' suggests obvious dangers (Swierstra et al. 2009). Jeff Carlson's 2007 novel *Plague Year* shows nanomachines created to fight cancer inevitably malfunctioning. This leads to the destruction of most planetary life forms. The concern is that:

> Just as genetic science appeared to open 'the book of life,' nanotechnology appears to give us an instruction manual for basic substances. Perhaps what makes nanotechnology seem to need nanoethics, then, is that it prompts questions about our control over nature – analogous to the ethical questions about changing human nature. (Kaebnick 2007)

One novel that neatly encapsulates the issue of runaway technology and human enhancement is John Robert Marlow's 2004 novel *Nano*, which offers the suggestion that enhancement must parallel nanotechnology – in short, we need to be smarter if we are to control such dangerous technology. Nanotechnology is released upon the world as a weapon that disassembles all in its path, a clear nod to Drexler's 'grey goo' scenario (Marlow 2004). Marlow makes the point that humankind is not evolutionarily advanced enough to deal with this new technology capable of being used for good and evil, and thus we will usually end up with evil. Molecular assemblers can create mature trees in seconds and free a dying bird from an oil spill through hydrocarbon disassembly, but nanotechnology is equally, in Marlow's protagonist's words, power: 'Invincibility. Immortality. Wealth from nothing' (Marlow 2004).

Marlow (2004) offers the familiar 'technology as too much power' scenario, and the reader awaits the inevitable demonstration of hubristic destruction for 'complex technologies create totalitarian technocracies'. The only way to avert doomsday is for the protagonist and his partner to inject nanites into their brain in hope of hyperevolution. This would give them the ability to come up with an answer to the question of how to stop the nanoswarm. The answer to militaristic totalitarianism is to allow the world to be run by two new nanoenhanced superintelligences – nanoelitism replacing nanodestruction. Or, a more positive message might be that enhancement relating to the evolution of one's moral capacity is required in using any other nanotech.

As Marlow's novel suggests, the idea of nanotechnology being used as an enhancement weapon is a popular one in science fiction, though we might remain sceptical of it being developed in the real world.

One specific area of concern is that of militarily directed nanoenhancement. The 2009 report on the ethics of human enhancement prepared for the US National Science Foundation identified several potential issues for ethical debate, including that of an enhanced elite being unfairly advantaged, particularly in military contexts (Allhoff et al. 2009). Joy's (2000) 'Faustian bargain', the trading of security for destructive technology, is difficult to pin down in factual terms due to a lack of public information about defence projects, particularly in China. We know that in the USA, MIT's Institute for Soldier Nanotechnologies (ISN) was established in 2002 with $50 million of funding from the Defense Advanced Research Projects Agency (DARPA). It aims to create a twenty-first century battle suit that 'combines high-tech capabilities with light weight and comfort…that monitors health, eases injuries, communicates automatically, and maybe even lends superhuman abilities' (Costandi 2006). Whether this is merely a better piece of battle gear, rather than enhancement in the strict sense of changing human physiology is a question made more complicated by simultaneous work being carried out on internal enhancement. This means the use of smart drugs to improve combat readiness as well as to create a defence against biological, bacteriological, or chemical combat agents. The drug modafinil is a prominent example, allegedly providing combatants with 40 h of alertness.

Science fiction can conceive of even more insidious weapons, as in Linda Nagata's *Deception Well* (1997), one of the four novels in her 'Nanotech Succession' series. The protagonist is infected by a 'cult' nanovirus that exudes nano-created psychoactive enzymes that transform anyone into adoring followers. He attempts to lead his followers to salvation; however, the reader is left to ask whether this is truly for some 'greater good' or if it is another exercise of power. Nagata's *Tech-Heaven* (1995) looks at the idea of power in a different but similarly disturbingly religious context. Here, nanobiotechnology is used to 'resurrect' the dead and so the protagonist has her husband's body frozen in the hope that nanomachines might 'raise the dead' and 'make the universe her playground' (Nagata 1995).

The idea that enhancement is dangerous because it might create human beings with 'superpowers' has been developed by some writers into depictions of societal dangers, such as elitism. The ongoing debate about the demographic shift to a large elderly population, a problem exacerbated by some countries' falling birth rates, as well as the rising costs of care, is dealt with by science fiction writers in more extreme terms. 'Immortality' treatments in dystopian texts, be they nanomanufactured or else biotechnological, are usually reserved for an elite class of society. James Gunn's *The Immortals* (1958) and Elizabeth Moon's *The Serrano Legacy* series (1993–2000) demonstrate this problem well; immortality markedly increases the gap between those who have access to treatment and those denied the privilege. Although we might consider current research into a nanoparticle anticancer drug as a wonderful, ethical way of improving the quality and duration of life (Selgelid 2009), will such be shared with all, given the cost of developing such treatments and the result of prior debates on the cost of retroviral medication?[9] Or are we back to the idea of nanoelitism?

The idea of a race of 'superhumans' with a godlike ability to manipulate matter may seem rather far-fetched. Nevertheless, it is an idea that adds to the ongoing debates in bioethics on the nature of being human. Elaine Graham (2002), looking at digital, cybernetic, and biomedical advances in terms of their impact on our understanding of human beings, argues that technology is seen as a kind of liberation. Humans are released from 'vulnerability, contingency and specificity' revealing 'a doctrine of humanity informed fundamentally by a distrust of the body, death and finitude' (Graham 2002). Graham contends that liberation results from the removal of any reminders of our limitations. This might include human flesh, which can reveal our weakness, frailty, and finitude. Yet for many science fiction writers, any superpowers inevitably lead to hubris and consequent disaster. In their view, humans should not 'play God'.

Perhaps an ethical distinction should be drawn between *overcoming* and *enhancing*.[10] Any medical procedure could conceivably change one's view of self, if only from sick or incapable to that of well and filled with potential. But a distinction between therapy and enablement (i.e. between Foucault's subject 'constructed within a benevolent narrative of amelioration and healing' (Lacombe 1996) and that of the subject pursuing perfectionism (Evans 2007)) frequently informs the debate. Critics of enhancement tend to stress the 'special' gift or uniqueness of human nature, asserting that any tampering with it would detract from our humanness and so blur the definition of 'humanity' (Kass 2003; Hughes 2007).

---

[9] Bawa (2005) has identified several barriers to commercial nanomedicine, such as high production costs and unclear regulatory guidelines. See also Allhoff (2009) and Freitas (2007).

[10] See The President's Council on Bioethics (2004) report *Beyond Therapy: Biotechnology and the Pursuit of Happiness.*

The most prevalent form of human enhancement in Western science fiction relates to what is called 'NBIC' convergence. NBIC, as the acronym has it, refers to the convergence of the disciplines of nanotechnology, biology, IT, and cognitive science. At one end of the spectrum, it involves uploading consciousness into a computer, while at the other it leads to nanotechnological engineering at neural levels, enhancing cognition or consciousness through neural implants or some kind of brain computer. Swiestra et al. (2009) have utilised the convergence issue to interrogate whether existing interpretative or cognitive frameworks would become deficient compared with enhanced nano-consciousness. Dupuy (2007) goes further to say that as cognitive science takes the leading role in NBIC, it effectively becomes a research programme examining new methods of perceiving reality.

Yet what might be lost if humanity were to evolve into cyborgism, becoming heavily reliant on technology? One answer comes from Kathleen Goonan's *Queen City Jazz* (1994), the first nanotech novel. Noonan's 'nanotech cycle' of *Queen City Jazz, Mississippi Blues* (1997), the prequel *Crescent City Rhapsody* (2000), and *Light Music* (2002) examines the issue of threats to identity from nanotechnology. In *Queen City Jazz*, Goonan's futuristic Cincinnati is a place where huge bio-engineered bees carry information through the streets and enormous nanotech energy-producing flowers burst from the tops of strange buildings. It also depicts an 'NBIC' future, in which uploaded consciousnesses form part of a hive or group mind. The 'information nanos' designed to educate, and described as 'cheap and easy brain growth for the masses', lead to a future lacking in humanity, freedom, and individuality (Goonan 1994). The problem is not entirely technological in Goonan's novel; instead it is the result of a hubristic scientist's reach exceeding his grasp. The nanoarchitect's desire to keep his mother alive leads him to upload her into the city mind, creating a tyrannical queen bee who rules the city. The moral of the story is of course that humanity is probably not to be trusted with the tool of nanotech. Our emotions, which arguably make us human, may lead us into error. Thus, Goonan suggests that nanotechnology is merely a part of the journey of self-discovery and that with our greater knowledge both of self and of matter, the tools of nanotechnology might be entrusted to us. For Goonan (2008), nanotech acts as a 'metaphor for the power of thought, and for the power of language. This may sound odd, but it seems that the more we understand matter the better we understand ourselves'.

Nanotechnological enhancement may lead us to a new transhuman or posthuman state. The terms 'transhuman', 'posthuman', and 'extropian' are often used without particular distinction, but the latter tends to be a more political expression (the five principles of extropianism published by Max More, co-founder of the Extropy Institute in 1992 include a focus on anti-government interference). Transhumanists see the enhancement movement as a unified effort between science and technology towards human progress, enhancement resulting in greater good. Transhumanism and posthumanism can be distinguished as stages on the enhancement path, as suggested by Nick Bostrom (2001), founder of the World Transhumanist Association. He defines 'posthuman' as a term for the more advanced beings that humans may one day design themselves into 'if we manage to upgrade our current human nature and radically extend our capacities'.

Some science fiction writers consider the idea of physical enhancement as a positive one, as Charles Sheffield does in *The Cyborg from Earth* (1998). In this tale, the protagonist is saved from death by nanomeds that remake his body. These are benign tools that, despite remodelling the protagonist's plump self, do not change his basic design. This is rather different to the posthumanist vision often explored by science fiction writers in terms of the 'postnatural' state of NBIC convergence, or uploaded consciousness (Thacker 2004). In this context, the human being might be discussed in terms more familiar to software engineers, as data, or as a composition of manipulated computer codes expressive of the equivalence between materiality and informatics. Human ontology might thus become digitised with the unification of flesh and data, bodies and information.[11] Venkatesan (2010) argues that NBIC convergence may imply the reductionist creation of a nanoself losing the 'considerations of spirituality, sociality, psychological and mental well-being, and the human needs for nurturing and sustenance' in 'the grandeur of the technological visions of convergence'.

Charles Stross' *Glasshouse* (2006) depicts a universe in which nanotechnology allows minds to be fully digitised, backed up, and restored; thus people are able to swap physical bodies and edit or manufacture memories at any time. His novel outlines the freedom that can come to a society with unlimited plasticity of identity, though his protagonists become paradoxically trapped in a terrifying social experiment, or panoptical 'glasshouse'.[12] In his collection *Accelerando* (2005) he repeats this negative stance on nano-induced freedom, arguing that enhancement may lead to slavery, not superhuman abilities. Stross (2005) offers a (deliberately sentimental) example of enhancement resulting in 'slavery' – kittens' neural networks enhanced and enslaved to missile guidance systems.

The theme of man's enslavement to his own technology returns us to the 'runaway doom' scenario, as in Linda Nagata's *The Bohr Maker* (1995). Nagata looks at nano as a potentially liberating technology; the heroine Phousita's horror at being internally colonised by nanotech is balanced by the power that it gives her to flee her poverty-stricken existence and browbeating husband. However, such a vision is contradicted by an alternative use of nanotech, one requiring a trade-off in terms of individual freedom, as characters become absorbed into a form of group mind. Bohr's maker is the catalyst for revolution, for political liberty, yet not liberty as we might relish it. This ambiguous novel shows that the consequences of nanotechnology's transformation of humanity are as yet not fully understood.

Stross' (2005) *Accelerando* collection introduces a debate on a new legal concept of 'what it is to be a person', namely, one 'that can cope with sentient corporations, artificial stupidities, secessionists from group minds, and reincarnated

---

[11] See Doyle's (1997) description of 'post-vital' molecular biology as based on the manipulation of codes in terms that sound familiar to software engineers.

[12] 'Charles Stross talks to io9 (2008a, b) about Sex, Prison, and Politics' at http://io9.com/342235/charles-stross-talks-to-io9-about-sex-prison-and-politics.

uploads'. Questioning what enhancement is *for*, Stross describes a group of 'uplifted' virtual lobsters who desire freedom from the world which has given them sentience without purpose or clarity.[13] Stross discusses the 'not human' economic system that a digital new world might construct, as well as the issue of dealing with a distributed 'superego' that occurs when one's self is uploaded to several technological devices, a nanobiology-enabled dissolution of the human into a posthuman distributed network, potentially resulting in a new form of group consciousness, or otherwise leading to utter fragmentation and loss.[14] Enhancement may create wonderful new forms of identity, or it could result in the complete destruction of our humanity. If being human is what gives us purpose, then another negative consequence of immortality is the pointlessness of eternal life. In texts that depict a NBIC future, when 'computer-based life will supplant biological life', as in Robert J. Sawyer's *Flashforward* (Sawyer 1999) or Greg Egan's *Permutation City* (1994), machine-human consciousnesses struggles to find a purpose. This is the major theme of Kage Baker's *Company* series (1997–2007), in which nano-repaired immortals (cyborgs) are sent back in time as historical scavengers, but are unable to find meaning in their lengthened existence.

In conclusion, Western science fiction writers have identified many of the topics that appear in journal articles on the social and ethical implications of nanotechnology, such as:

- Whether nanomedicine is beneficial or leads to enhancement with potentially negative effects in terms of creating new elite and 'inhuman' identities
- Whether nanotechnology can be uncontrollably destructive
- Whether it will measurably increase our loss of privacy

What other nanoissues can be identified, outside of the pages of science fiction novels? And to return to our earlier question, do the East and West have varying perceptions of nanoissues? If so, how have nanoethics debates developed in the EU and in China?

2.2.2 Nanoethics Debates in the EU and in China

Either the ethics of NT (nanotechnology) will catch up or the science will slow down. (Mnyusiwalla et al. 2003)

Bioethics does not serve society well simply by promoting a respect for other cultures. That's nice, but not enough. It better promotes society by helping to develop the criteria and standards for knowing which practices and values should be accepted and affirmed, which simply tolerated, and which rejected. (Callahan 2000)

---

[13] For comment on the collection, see Lou Anders, 'New Directions: Decoding the Imagination of Charles Stross. An Interview', at http://www.infinityplus.co.uk/nonfiction/intcs.htm.

[14] As popular in the many science fiction texts that reference Pierre Teilhard de Chardin's idea of the noosphere, or next evolutionary, collective level of consciousness; is a term used by Stross in the *Accelerando* collection.

The mainstreaming of debates about nanotechnology in the West is due in part to the publication of reports by Greenpeace, the Royal Society and Royal Academy (RS/RAE), and insurance company Swiss Re in 2003–2004. The first of these called for greater public debate so as to ensure the environmental benefits and risks of nanotechnology were understood, the second looked at chemical regulation and general socio-ethical issues, and the third looked at nanoparticle risks.

One platform for nanoethics debates in the West is the journal *NanoEthics: Ethics for Technologies that Converge at the Nanoscale* (Springer), begun in 2007. Of course, literature on nanoethics appeared well before that date, and Kjolberg and Wickson have written a useful account of pre-2007 literature on the nanoethics field (1994–2006), demonstrating a peak in publications during 2003. Their database reveals four key areas of interest in the EU and USA. First, governance, the 'processes and institutions for decision making, regulation, legislation and public engagement'. (Kjølberg and Wickson 2007). Early discussion of the ethics of nanotechnology in the EU and USA has understandably tended to focus on risks to human health and to the environment, a context that places an emphasis on determining whether precautionary and regulatory measures are appropriate and efficient. The other three areas of particular interest defined by Kjølberg and Wickson (2007) were:

> *Perception*: examining how nano is understood, presented, talked about and imaged. *Science*: exploring the practice of nano S&T [Science & Technology] development and instrumentation. *Philosophy*: engaging questions of metaphysics, the natural/artificial and ethical norms.

The issue of risk governance due to fears of toxicity will be dealt with later in this book, as the differing approaches of the EU and China to nanopolicy may indicate some basis for further comparison. Before looking at those areas of risk policy and public perception, however, we need to first establish the other concerns that exist. If there is broad agreement on these issues, then we should look at whether the difference then consists in how these issues are prioritised and handled for it is important to determine which socio-ethical framework forms the context for discussion. Once we have identified the core values that appear to underpin each region's general approach to ethics, we may be able to suggest a way of reaching global consensus. To this end, we are interested in looking at any differences between the 'narratives' of each region and the way that these contextualise nanotechnology. This might mean understanding nanotechnology as economic and scientific progress or as part of a debate over precaution and responsibility. To identify these contexts, this book will focus on:

- Critical literature on nanoethics published in each region
- The regional bioethics background already in existence

2.3   Debates on Nanoethics – EU

The 2008 report from the Rathenau Institute in the Netherlands (which promotes the formation of public and political opinion on S&T) examined a list of societal questions about nanotechnology compiled chiefly by the EU, to 'determine whether this list is complete, and to establish the degree of urgency which the NGOs attach to various issues' (Hanssen et al. 2008). The conclusion was that debates on nanotechnology, given the breadth of the area, need to be targeted, clearly defined, and considered separately from other debates on health and S&T. The report also notes current public awareness of nanotechnology as 'extremely low' (Hanssen et al. 2008).

Western commentators on nanotechnology issues usually come up with a list that, apart from toxicity related primarily to health and environmental risks, tends to incorporate the following:

- Privacy
- Enhancement
- Military usage
- Equity, including the equity issue in developing countries
- Public involvement (Wood et al. 2008)

Other issues are more specific, such as Spagnolo and Daloiso's (2009) suggestion that one area of ethical concern is the shift from patient-doctor interaction to home-care technology used by the patient (portable medical nanotechnology). Schummer (2007) has also raised the issue of intellectual property rights. Another concern, less widely discussed, has been that of technological determinism. The vision of 'autonomous' technology, with the human being reduced to a spectator of progress, assumes a pessimistic resignation to deterministic forces, allowing technology to act as a force unto itself, with neither the public nor scientists claiming agency over it. Jamison (2009) has taken a view that the strong economic drive may cause a S&T 'hegemonic' narrative to appear, in terms of which there is little concern for societal values. He argues that nanotechnology is developing according to a commercial model; this does not mean that its applications cannot have social value, but this value would inevitably be only a secondary concern.

How best to categorise all these issues? At this point it is useful to introduce Barakat and Jaio's (2010) three categories of nanoethical issues:

1. Life-basics ethics (risk and 'first do no harm' ethics). This includes concepts like autonomy, military applications, fear of uncontrolled actions (e.g. runaway reactions and uncontrolled self replications), and health hazards.
2. Life-quality ethics (justice and equality ethics). This includes ideas like the nanodivide where the gap between rich and poor nations will increase.
3. Life and human definition ethics. This includes the concept of integrity and issues related to human change (Barakat and Jiao).

The journal *NanoEthics* announced its mission as that of focusing on issues that 'include individual health, wellbeing and human enhancement, human integrity and autonomy, distribution of the costs and benefits, threats to culture and tradition and to political and economic stability'.[15] The years 2007–2010 covered general areas of ongoing concern, with articles on risk and regulation, enhancement, nanomedicine, NBIC, and public perceptions – in addition to other topics including justice, teaching nanoethics, green nanotechnology, intellectual property rights, ambivalence towards technology, Australian and Thai nanoresearch, and nanoethics.

What of more recent topics? The journal has addressed the controversial issue of animal disenhancement – an interesting branch of the nanoenhancement debate (volume 6, April 2012).[16] A series on 'imaging' the nanoscale raised the question of the relationship between science and art (volume 5, August 2011). Ruivenkamp and Rip's article (2011) on the speculative nature of nanoethics demonstrates how this problem is exacerbated by the essentially 'unseen' nature of the nanoscale. The same issue, via Grinbaum's (2011) article on the iconography of nanotechnology, picked up this issue of 'the unseen', describing the knowledge gap between scientist and layperson as a 'two-class system'.

This brings us to a nexus of issues centred on the role of the public. This topic, of public involvement in the nanodebate and in nanopolicy, will be discussed in more detail in Chapter Four, but the main aspects of the debate can be briefly noted here. The first is that of public education, more specifically noted by Grinbaum as the problem of communicating a complex, microscopic science that cannot be demonstrated easily. Science can be lexically challenging for a wider audience liable to lose patience with the debate if the issues are not particularly clear-cut, an educational difficulty confirmed by the ongoing debates on global warming which reveal how 'disinformation' through lobbying may frustrate public discussion (Collins 2007). In addition, risk reporting tends to take place in the general media at the expense of articles on the benefits of nanotechnology, which are confined to business or science sections with much smaller audiences (Stephens 2005). Moreover, nanotechnology risk has a clear sensationalist appeal, which if amplified by the media can lead to stigmatisation of a technology to the point that it becomes 'blemished' or 'tainted' by 'discourses of risk' (Petersen and Anderson 2007).[17]

The next issue is that articles on the undefined risks of a product backed by strong economic imperatives, with 'low gates' to the market in regulatory terms, are symptomatic of a continued interest in 'anticipatory assessment' as well as perceived tensions between public trust and corporate responsibility (Robison 2011). Some of the more interesting debates in *NanoEthics* have been on the 'decentralisa-

---

[15] The focus of the journal is described at

http://www.springer.com/social+sciences/applied+ethics/journal/11569. The statement includes: 'additionally there are meta-issues including the neutrality or otherwise of technology, designing technology in a value-sensitive way, and the control of scientific research'.

[16] For example, chickens might be bred blind (purposely disenhanced) to improve their welfare in animal commodity contexts such as dark breeding pens.

[17] See also Russell (2010), Irwin (2001), and Anderson et al. (2009).

tion' of nanoresearch and policy, for instance, by shifting the emphasis onto the person rather than the institution or policymaking body.[18] On this topic, Nielsen et al. (2011) claim that the nanotechnology field 'went global before it had reached a mature state' and that there 'has not been, as it were, a centre delivering core knowledge to be consumed, imitated, opposed or modulated in the periphery'. Is it therefore up to the scientist and the public to take responsibility? Perhaps philosopher Han Jonas's familiar notion – that we need an ethics of responsibility now that power has outstripped knowledge in terms of technology – has resurfaced.

2.4   Debates on Nanoethics: China

While China has given the risk issue priority in cases such as the credible effects of nanoparticles on humans and the environment, there has also been acknowledgement of other societal concerns. This is mentioned in The Director's Note in the 2007–2008 Annual Report of the China Nanosafety Lab, in which it is stated that the lab 'must take an extensive and deep research of nanotechnological influence on human health, environment, and social problems' (Zhao 2007–2008). The problem is that despite mentioning these social problems, the report does not provide any further detail on what they might be. This is explained by the fact that concerns about nanoethics started later in China than in the EU. China's belief in social progress through scientific development means that economic impetus, rather than societal concern, is often the major driver and overriding impetus behind new technologies. Thus, compared to the reasonably extensive debate in the EU, the Chinese nanoethics debate is fairly low key.

Ying (2006) likens nanotechnology concerns to genetically enhanced (GM) food issues, implying a public acceptance issue, albeit without much exploration of the problem, a topic that will be discussed in more detail in Chap. 2. Li (2010) discusses environmental problems by giving the example of a Korean company that halted production of a particular model of washing machine following pressure from Friends of the Earth, as well as noting wider issues such as increased lifespan due to nanomedical development and the societal impact that this might incur. Wang (2010) notes potential problems related to consumer rights (in the context of food and cosmetics products), in addition to privacy and intellectual property rights issues. Fan (2010), perhaps the most 'Western' in his strong societal emphasis on nanoethics, notes that 'compared with safety issues, research on ethical, legal and social issues should be strengthened', as should dialogue between the scientific community and the public.

But as Cao and Li (2010) note, the shift from an approach to nanotechnology policy and regulation based predominantly in the scientific community, to one founded on regulations developed in collaboration with social scientists is an

---

[18] See, for example, Myska (2011) on the Norwegian NANOTRUST initiative, focusing on the trustworthiness of researchers.

ongoing process. Choi's 2003 study listed the following issues which need to be addressed by Asian nanoethicists:

- Equity between those with access to technology and those without, both in terms of developed versus underdeveloped countries and internally within rural and urban populations
- Privacy issues in Confucian systems (there has long been debate over the individual's right to privacy of medical information, e.g. particularly in Japan)
- Gender issues (as the majority of nanoscientists are male)
- Brain implants and other issues relating to human enhancement
- Undue inducement, for human subjects in nanomedical clinical trials, for example
- Military uses of new technologies
- Environmental toxicity, including how effectively nanowaste can be managed in space-limited countries with large populations.

This list is not markedly different to Western concerns; however, we can more fully appreciate the differences by comparing China's approach with that of Taiwan. Taiwan's National Strategic Plan for Responsible Nanotechnology reflects the government's ambition to realise the full potential of nanotechnologies while acknowledging that there may be a harmful societal impact. This outlook on technology policy can be summed up in the following section from the most recent Taiwanese Science and Technology Development Plan (2009–2012):

> Appropriate ethical and legal responses may be needed to deal with the risks posed by new technologies to life and the environment, and the ethical conflicts they cause…Unlike such areas as medical biotechnology, where ethics committees have been established, little has been done thus far to address research ethics in many new technological fields in Taiwan (such as genetic technology and nanotechnology). (National Science Council 2009)[19]

The Thai government is currently reviewing the country's first strategy plan on nanotechnology safety and ethics, drafted in 2011. This sounds more proactive in terms of nanoethics than China, although Harmon et al. (2011) note that very little ethical debate has really taken place in Taiwan. Despite the formation of Taiwan's National Science and Technology Programme for Nanoscience and Nanotechnology, an organisation that coordinates various regulatory bodies, there are no nanoscience ethics committees, and it is arguable that civil society groups have been marginalised. Given the so-called Asian drive towards technology as economically advantageous, the Taiwanese nanonarrative, like that of China, tends to be positive and 'serves in some respects as a cultural counter-point' to the more cautionary approach more often found in Europe (Harmon et al. 2011).

In conclusion, the above discussion indicates that there are several familiar topics in nanoethics debates that resurface frequently. In both China and the EU, the issue of risk to human health and environment is clearly a priority, the enhancement, military, and privacy debates less so, while the equity issue tends to be even further down the scale. The table below summarises the general approach to such issues in both the EU and China (Table 2.1).

---

[19] See Purra and Richmond (2010).

Table 2.1 Main nanoethics issues in China and in the EU

Enhancement – Pro (West)	Enhancement – Con (West)
A 'better human' with increased longevity, fewer health issues, possibly even enhanced intelligence	New elites (enhancement only for the rich); loss of 'humanness'; increased longevity, i.e. more pressure on resources
Enhancement – Pro (East)	**Enhancement – Con (East)**
Not publically debated	*Not publically debated*
Health – Pro (West)	**Health – Con (West)**
Better drugs; portability (better health care for remote and rural communities); economic benefit of reduced health-care spending and innovation	Threats to human health through dermal exposure or inhalation or ingestion of nanoparticles
Health – Pro (East)	**Health – Con (East)**
Health-care improvements, commercial advantages	Threats to human health through dermal exposure or inhalation or ingestion of nanoparticles
Environment – Pro (West)	**Environment – Con (West)**
Amelioration of contaminated/non-potable groundwater; increased agricultural yields	Potential threat to environmental and so to human health
Environment – Pro (East)	**Environment – Con (East)**
Improved fertilisers, thus increased agricultural yields	Potential threat to environmental health; nanowaste and space limitations
Military applications – Pro (West)	**Military applications – Con (West)**
Better wound care; more precise targeting (less collateral damage); better weaponry	'Unequal wars'; creation of 'supersoldiers'
Military applications – Pro (East)	**Military applications – Con (East)**
Better weaponry, often with commercial value	
Equity – Pro (West)	**Equity – Con (West)**
General economic boost, particularly to the construction, energy, medtech, and IT industries	Increased (nano)divide between developed and developing countries – issues of distributive justice, global benefit
Equity – Pro (East)	**Equity – Con (East)**
General economic benefit globally in terms of new products	Nanodivides within Asian countries with large rural and/or poor populations; undue inducement
Privacy – Pro (West)	**Privacy – Con (West)**
Smaller and less obtrusive surveillance devices; greater national security	Increasingly miniaturised surveillance devices lead to potential loss of civil liberties, privacy of medical information becomes an issue
Privacy – Pro (East)	**Privacy – Con (East)**
Greater national security	Access to medical records can be an issue
Public perception – Pro (West)	**Public perception – Con (West)**
	Public acceptance vital for economic success of new products
Significance of public perception – Pro (East)	**Public perception – Con (East)**
Stimulus to purchasing power; 'nano' is advertising plus advancement	Public fears might impede economic progress

References

Adams W, Williams L (2006) Nanotechnology demystified. McGraw-Hill Professional Publishing, Blacklick

Allhoff F (2007) On the autonomy and justification of nanoethics. NanoEthics 1:185–210

Allhoff F (2009) The coming era of nanomedicine. Am J Bioeth 9(10):3–11

Allhoff F, Lin P, Weckert J, Moor J (2009) Ethics of human enhancement: 25 questions & answers. Available via US National Science Foundation. http://www.humanenhance.com/NSF_report.pdf. Accessed 21 Dec 2014

Allhoff F, Lin P, Moore D (2010) What is nanotechnology and why does it matter? From science to ethics. Wiley-Blackwell, Chichester

Anderson A, Petersen A, Wilkinson C, Allan S (2009) Nanotechnology, risk and communication. Palgrave Macmillan, Basingstoke

Arnall A (2003) Future technologies, today's choices. Available via Greenpeace Environmental Trust. http://www.greenpeace.org.uk/MultimediaFiles/Live/FullReport/5886.pdf. Accessed 20 Dec 2014

Barakat N, Jiao H (2010) Proposed strategies for teaching ethics of nanotechnology. NanoEthics 4:221–228

Bawa R (2005) Will the nanomedicine patent 'land grab' thwart commercialization? Nanomedicine 1(4):346–350

Bawa R, Johnson S (2008) Emerging issues in nanomedicine and ethics. In: Allhoff F, Lin P (eds) Nanotechnology and society: current and emerging social and ethical issues. Springer, Dordrecht, pp 207–223

BBC News (2007) Russia tests giant fuel-air bomb, 12 September. http://news.bbc.co.uk/2/hi/europe/6990815.stm. Accessed 21 Dec 2014

Belt H (2009) Playing God in Frankenstein's footsteps: synthetic biology and the meaning of life. NanoEthics 3(3):257–268

Bennett-Woods D (2008) Nanotechnology: ethics and society. CRC Press, New York

Berger M (2007) Debunking the trillion dollar nanotechnology market size hype. Available via Nanowerk. http://www.nanowerk.com/spotlight/spotid=1792.php. Accessed 21 Dec 2014

Berne RW (2004) Towards the conscientious development of ethical nanotechnology. Eng Ethics 10:627–638

Berne R (2005) Nanotalk: conversations with scientists and engineers about ethics, meaning, and belief in the development of nanotechnology. Lawrence Erlbaum, Mahwah

Berne R (2008) Science fiction, nano-ethics and the moral imagination. In: Fisher E, Selin C, Wetmore J (eds) The yearbook of nanotechnology in society, vol I. Springer, New York, pp 291–302

Bostrom N (2001) Transhumanist values. http://www.nickbostrom.com/tra/values.html. Accessed 21 Dec 2014

Bostrom N (2007) Technological revolutions: ethics and policy in the dark. In: de S Cameron NM, Ellen Mitchell M (eds) Nanoscale: issues and perspectives for the nano century. Wiley, Hoboken, pp 129–152

Callahan D (2000) Universalism and particularism: fighting to a draw. Hastings Centre Rep 30(1):37–44

Cao N, Li S (2010) The development of China's nanotechnology needs allies in humanities and social sciences. Chinese Social Sciences Today, 25 September. http://sspress.cass.cn/newspaper/paper.aspx?Id=1000129. Accessed 21 Dec 2014

Centre for Responsible Nanotechnology (2002–2008) What is nanotechnology? http://www.crnano.org/whatis.htm. Accessed 21 Dec 2014

Changsorn P (2004) Firms see lower costs, more profit in nanotech. The Nation (Thailand), 22 November, p unknown

Chen S (2007) Back to the future. China Daily, 30 August 2007, p 18, http://www.chinadaily.com.cn/cndy/2007-08/30/content_6066566.htm. Accessed 21 Dec 2014

Choi K (2003) Ethical issues of nanotechnology development in the Asia-Pacific region. In: Regional meeting on the ethics of science and technology. UNESCO, Bangkok, pp 327–376

Cobb M, Macoubrie J (2004) Public perceptions about nanotechnology. J Nanoparticle Res 6:395–405

Collins JC (2007) Nanotechnology and society: a call for rational dialogue. In: de S. Cameron NM, Ellen Mitchell M (eds) Nanoscale: issues and perspectives for nano century. Wiley, Hoboken, pp 115–128

Costandi M (2006) The 'Nano-enhanced super soldier'. http://neurophilosophy.wordpress.com/2006/07/09/the-nano-enhanced-super-soldier. Accessed 21 Dec 2014

Davies S, Macnaghten P (2010) Narratives of mastery and resistance: lay ethics of nanotechnology. NanoEthics 4:141–151

de S Cameron NM (2007) Towards nanoethics? In: de S Cameron NM, Ellen Mitchell M (eds) Nanoscale: issues and perspectives for nano century. Wiley, Hoboken, pp 281–294

Decker M, Li Z (2009) Dealing with nanoparticles: a comparison between Chinese and European approaches to nanotechnology. In: Ladikas M (ed) Embedding society in science & technology policy. European and Chinese perspectives. European Commission, Brussels, pp 91–123. http://ec.europa.eu/research/science-society/document_library/pdf_06/european-and-chinese-perspectives_en.pdf. Accessed 21 Dec 2014

Dillow C (2011) The first self-powering nano-device that can also transmit wireless data. Popsci, 17 June. http://www.popsci.com/science/article/2011-06/first-self-powering-nano-device-can-also-transmit-data-over-long-distances. Accessed 21 Dec 2014

Doubleday R (2007) Risk, public engagement and reflexivity: alternative framings of the public dimensions of nanotechnology. Health Risk Soc 9(2):211–227

Doyle R (1997) On beyond living: rhetorical transformations of the life sciences. Stanford University Press, Stanford

Dunn J (2004) A mini revolution. http://www.foodmanufacture.co.uk/news/fullstory.php/aid/472/A_mini_revolution.html. Accessed 21 Dec 2014

Dupuy J (2007) Some pitfalls in the philosophical foundations of nanoethics. J Med Philos 32:237–261

Dupuy J (2010) The narratology of lay ethics. NanoEthics 4:153–170

Dupuy J, Grinbaum A (2006) Living with uncertainty: towards the ongoing normative assessment of nanotechnology. In: Schummer J, Baird D (eds) Nanotechnology challenges – implications for philosophy, ethics and society. World Scientific Publishing, Singapore, pp 287–314

ETC Group (2010) The big downturn? Nanogeopolitics. http://www.etcgroup.org/sites/www.etcgroup.org/files/publication/pdf_file/nano_big4web.pdf. Accessed 21 Dec 2014

European Commission (2004) Communication from the Commission: towards a European strategy for nanotechnology. ftp://ftp.cordis.europa.eu/pub/nanotechnology/docs/nano_com_en_new.pdf. Accessed 21 Dec 2014

Evans D (2007) Ethics, nanotechnology and health. In: Tenhave H (ed) Nanotechnologies, ethics, and politics. UNESCO, Paris, pp 125–154

Fan C (2010) The ethical environment of nano-science. Paper presented at the 3rd international workshop on innovation and performance management, University of Kent, 1–4 July 2010

Ferrari A (2010) Developments in the debate on nanoethics: traditional approaches and the need for new kinds of analysis. NanoEthics 4:27–52

Feynman R (1960) There's plenty of room at the bottom. Eng Sci 23(5):22–36

Fisher F (2005) Lessons learned from the ethical, legal & social implications program (ELSI): planning a societal implications research program for the national nanotechnology program. Technol Soc 27:321–328

Freitas R (1998–2006) Nanomedicine. Available via Foresight Institute. http://www.foresight.org/Nanomedicine/. Accessed 21 Dec 2014

Freitas R (2007) Personal choice in the coming era of nanomedicine. In: Lin P, Moor J, Weckert J, Allhoff F (eds) Nanoethics: the ethical and social implications of nanotechnology. Wiley, Hoboken, pp 161–172

Gasman L (2006) Nanotechnology applications and markets. Artech House, Norwood

Goonan KA (1994) Queen city Jazz. Tom Doherty Associates, New York

Gordijn B (2005) Nanoethics: from utopian dreams and apocalyptic nightmares. Sci Eng Ethics 11:521–533

Gorman ME, Groves JF, Shrager J, Baird D, Schummer J (2007) Societal dimensions of nanotechnology as training zone: results from a pilot project. In: Baird D, Nordmann A, Schummer J (eds) Discovering the nanoscale. Ios Press, Amsterdam, pp 63–73

Graham E (2002) Representations of the post/human. Monsters, aliens and others in popular culture. Manchester University Press, Manchester

Grinbaum A (2011) Nanotechnological icons. NanoEthics 5(2):195–202

Grunwald A (2010) From speculative nanoethics to explorative philosophy of nanotechnology. NanoEthics 4:91–101

Hanssen L, Walhout B, van Est R (2008) Ten lessons for a nanodialogue: the Dutch debate about nanotechnology thus far. Rathenau Institute, The Hague

Harmon SHE, Yen SY, Tang SM (2011) Invigorating nanoethics: recommendations for improving deliberations in Taiwan and beyond. NanoEthics 5(3):309–318

Helmut Kaiser Consultancy (2006) Nanotechnology in food and food processing worldwide, 2003–2006 – 2010–2015. http://www.hkc22.com/Nanofood.html. Accessed 21 Dec 2014

Hook C (2007) Nanotechnology and the future of medicine. In: Baird D, Nordmann A, Schummer J (eds) Discovering the nanoscale. Ios Press, Amsterdam, pp 337–360

Howard C (2010) Nanowhere. Lykeion Books, Kindle for Android

Hughes J (2007) Beyond human nature: the debate over nanotechnological enhancement. In: de S Cameron NM, Ellen Mitchell M (eds) Nanoscale: issues and perspectives for nano century. Wiley, Hoboken, pp 61–70

Hwang D (2010) Ranking the nations on nanotech. Available via Solid State Technology. http://www.electroiq.com/articles/stm/2010/08/ranking-the-nations.html. Accessed 21 Dec 2014

i09 (2008a) io9 talks to Kathleen Ann Goonan about Nanopunk and Jazz. Available via Nanowerk. http://www.nanowerk.com/news/newsid=4123.php. Accessed 21 Dec 2014

i09 (2008b) Charles Stross talks to io9 about sex, prison, and politics. http://io9.com/342235/charles-stross-talks-to-io9-about-sex-prison-and-politics. Accessed 21 Dec 2014

Irwin A (2001) Constructing the scientific citizen: science and democracy in the biosciences. Public Underst Sci 10:1–18

Jamison A (2009) Can nanotechnology be just? On nanotechnology and the emerging movement for global justice. NanoEthics 3:129–136

Johnson DG (2007) Ethics and technology 'in the making': an essay on the challenge of nanoethics. NanoEthics 1:21–30

Johnson K (2010) Advances in nanotechnology hold huge potential promise in upstream applications. Available via The American Oil & Gas Reporter. http://www.beg.utexas.edu/aec/pdf/0710_AEC_Eprint.pdf. Accessed 21 Dec 2014

Joseph T, Morrison M (2006) Nanoforum report: nanotechnology in agriculture and food. ftp://ftp.cordis.europa.eu/pub/nanotechnology/docs/nanotechnology_in_agriculture_and_food.pdf. Accessed 21 Dec 2014

Joy B (2000) Why the future doesn't need us. Available via Wired. http://www.wired.com/wired/archive/8.04/joy_pr.html. Accessed 21 Dec 2014

Kaebnick GE (2007) Field talk. Hastings Centre Rep 37(1):14–17

Kass L (2003) Ageless bodies, happy souls: biotechnology and the pursuit of perfection. New Atlantis 1(Spring):9–28

Keiper A (2007) Nanoethics as a discipline? New Atlantis 16(Spring):55–67

Kimbrell GA (2007) The potential environmental hazards of nanotechnology and the applicability of existing law. In: de S Cameron NM, Ellen Mitchell M (eds) Nanoscale: issues and perspectives for nano century. Wiley, Hoboken, pp 213–214

Kjølberg K, Wickson F (2007) Social and ethical interactions with nano: mapping the early literature. NanoEthics 1:89–104

Koepsell D (2010) Scientists' moral responsibility and dangerous technology R&D. Sci Eng Ethics 16(1):119–133

Kuzma J, VerHage P (2006) Nanotechnology in agriculture and food production: anticipated applications. http://www.nanotechproject.org/file_download/files/PEN4_AgFood.pdf. Accessed 21 Dec 2014

Lacombe D (1996) Reforming Foucault: a critique of the social control thesis. Br J Sociol 47(2):332–352

Lewenstein B (2005) What counts as a "Social and ethical issue" in nanotechnology? Hyle 11(1):5–18

Li SH (2010) Small world, big results. Chinese Social Science News, 8 April, pp 1–2

Li Y, Somorjai G (2010) Nanoscale advances in catalysis and energy applications. NanoLett 10(7):2289–2295

Lin P, Moor J, Weckert J (2007) Nanoscience and nanoethics: defining the disciplines. In: Allhoff F (ed) Nanoethics: the ethical and social implications of nanotechnology. Wiley, Hoboken, pp 3–16

Liu X, Zhang P, Li X, Chen H, Dang Y, Larson C, Roco M, Wang X (2009) Trends for nanotechnology development in China, Russia, and India. J Nanoparticle Res 11(8):1845–1866

López J (2004) Bridging the gaps: science fiction in nanotechnology. Hyle 10(2):129–152. http://www.hyle.org/journal/issues/10-2/lopez.htm. Accessed 21 Dec 2014

Maclurcan D (2005) Nanotechnology and developing countries – Part 2: What realities? http://www.azonano.com/article.aspx?ArticleID=1429. Accessed 19 Dec 2014

Marlow JR (2004) Nano. Forge Books, New York

Mehlman MJ (2003) Wondergenes: genetic enhancement and the future of society. Indiana University Press, Bloomington

Mehta M (2002) Privacy vs. surveillance: how to avoid a nano-panoptic future. Can Chem News 54(10):31–33

Mick J (2008) Great Buckyballs! Storing hydrogen with carbon nanostructures. http://www.dailytech.com/Great+Buckyballs++Storing+Hydrogen+with+Carbon+Nanostructures/article. Accessed 21 Dec 2014

Milburn C (2008) Nanovision: engineering the future. Duke University Press, Durham

Milner A (2006) Framing catastrophe. The problem of ending in dystopian fiction. In: Milner A, Ryan M, Savage R (eds) Imagining the future: Utopia and dystopia. Arena Publications, North Carlton, pp 333–356

Mitchell ME (2007) Scientific promise: reflections on nano-hype. In: de S. Cameron NM, Mitchell M Ellen (eds) Nanoscale: issues and perspectives for nano century. Wiley, Hoboken, pp 43–60

Mnyusiwalla A, Daar AS, Singer P (2003) Mind the gap: science and ethics in nanotechnology. Nanotechnology 14:R9–R13

Moor J, Weckert J (2004) Nanoethics: assessing the nanoscale from an ethical point of view. In: Baird D, Nordmann A, Schummer J (eds) Discovering the nanoscale. Ios Press, Amsterdam, pp 301–310

Myskja BK (2011) Trustworthy nanotechnology: risk, engagement and responsibility. NanoEthics 5(1):49–56

Nagata L (1995) Tech-Heaven, Kindle edn

Nano comes to Clifford Falls. http://escapepod.org/2006/10/12/ep075-nano-comes-to-clifford-falls/. Accessed 23 Dec 2014

Nanowerk (2009) Consumer products drop nanotechnology claims for fear of consumer recoil. http://www.nanowerk.com/news/newsid=11181.php. Accessed 19 Dec 2014

National Research Council (2003) Implications of Emerging Micro- and Nanotechnologies. National Academies Press, Washington, DC

National Science Council (2009) National science and technology development plan (2009–2012). http://www.most.gov.tw/public/Attachment/91214167571.PDF. Accessed 19 Dec 2014

Nature Nanotechnology (2007) Beware of big brother. http://www.nature.com/nnano/journal/v2/n1/full/nnano.2006.207.html. Accessed 19 Dec 2014

Nielsen KN, Åm TG, Nydal R (2011) Centre and periphery of nano – a Norwegian context. NanoEthics 5(1):87–98

Niu GA (2008) Techno-orientalism, nanotechnology, posthumans, and post-posthumans in Neal Stephenson's and Linda Nagata's science fiction. MELUS 33(4):73–96

Nordmann A (2005) Noumenal technology: reflections on the incredible tininess. http://scholar.lib.vt.edu/ejournals/SPT/v8n3/nordmann.html. Accessed 21 Dec 2014

Nordmann A (2006) No future for nanotechnology? Historical development vs global expansion. In: Bainbridge SW, Roco MC (eds) Progress in convergence: technologies for human wellbeing. Blackwell, Boston, pp 43–66

Nordmann A (2007) If and then: a critique of speculative nanoethics. NanoEthics 1:31–46

Nordmann A, Rip A (2009) Mind the gap revisited. Nat Nanotechnol 4:273–274

Pagliaro MP (2010) Nano-age: how nanotechnology changes our future. Wiley, Weinheim

Petersen A, Anderson A (2007) A question of balance or blind faith? Scientists' and science policymakers: representations of the benefits and risks of nanotechnologies. NanoEthics 1(3):243–256

Pravda.Ru (2010) Enterprise of nanotechnologies opens in Russia. http://english.pravda.ru/science/tech/28-04-2010/113140-nanotechnology-0/. Accessed 19 Dec 2014

Preston CJ, Sheinin MY, Sproat DJ, Swarup VP (2010) The novelty of nano and the regulatory challenge of newness. NanoEthics 4:13–26

Purra M, Richmond N (2010) Mapping emerging nanotechnology policies and regulation: the case of Taiwan. http://www.lse.ac.uk/internationalRelations/centresandunits/regulatingnanotechnologies/nanopdfs/Taiwan2010.pdf. Accessed 21 Dec 2014

Roache R (2008) Ethics, speculation, and values. NanoEthics 2:317–327

Robison W (2011) Nano-technology, ethics, and risks. NanoEthics 5(1):1–13

Rodgers P (2006) Nanoelectronics: single file. http://www.nature.com/nnano/reshigh/2006/0606/full/nnano.2006.5.html. Accessed 21 Dec 2014

Ruivenkamp M, Rip A (2011) Entanglement of imaging and imagining of nanotechnology. NanoEthics 5(2):185–193

Russell N (2010) Communicating science. Cambridge University Press, Cambridge

Sametband M, Shweky I, Banin U, Mandler D, Almog J (2005) Application of nanoparticles for the enhancement of latent fingerprints. Chem Commun 12(11):1142–1144

Savulescu J (2009) Genetic interventions and the ethics of enhancement of human beings. In: Kaplan DM (ed) Readings in the philosophy of technology. Rowman & Littlefield, Lanham, pp 417–430

Sawyer RL (1999) Flashforward. Tor Books, New York

Scheufele D, Lewenstein B (2005) The public and nanotechnology. J Nanoparticle Res 7:659–667

Schmid G, Brune H, Ernst H, Grunwald W, Grunwald A, Hoffmann H, Krug H, Janich P, Mayor M, Rathgeber W, Simon U, Vogel V, Wyrwa D (2006) Nanotechnology: assessment and perspectives, vol 27, pp 302–319

Schummer J (2007) Identifying ethical issues of nanotechnologies. In: Tenhave H (ed) Nanotechnologies, ethics, and politics. UNESCO, Paris, pp 79–99

Schwarz AE (2009) Green dreams of reason. Green nanotechnology between visions of excess and control. NanoEthics 3:109–118

Science Daily (2008) Nanotechnology may be used for food safety. Science Daily, 28 December. http://www.sciencedaily.com/releases/2008/12/081228194854.htm. Accessed 21 Dec 2014

Sebag-Montefiore C (2012) Cultural exchange: Chinese science fiction's subversive politics. Los Angeles Times, 25 March. http://articles.latimes.com/2012/mar/25/entertainment/la-ca-china-culture-20120325. Accessed 21 Dec 2014

Selgelid MJ (2009) Dual-use codes of conduct: lessons from the life sciences. NanoEthics 3:175–183

Spagnolo AG, Daloiso V (2009) Outlining ethical issues in nanotechnologies. Bioethics 23(7):394–402

Sputnik (2009) Russia to invest \$9.4 bln in nanotech projects in 2008–2015. http://en.rian.ru/russia/20090918/156174098.html. Accessed 23 Dec 2014

Stanton J (2006) Nanotechnology, surveillance, and society: issues and innovations for social research. In: Roco MC, Bainbridge WS (eds) Nanotechnology: societal implications – individual perspectives. Springer, Berlin, pp 87–97

Stephens LF (2005) News narratives about nano S&T in major U.S. and non-U.S. newspapers. Sci Commun 27(2):175–199

Stephenson N (1995) The diamond age. Penguin, London

Storrs Hall J (2005) Nanofuture: what's next for nanotechnology. Prometheus, New York

Stross C (2005) Accelerando. Ace, New York

Stross C, quoted in an interview with Sirius R. When I called Charlie Stross a Dirty name.... Transhumanist. Acceler8tor (sic) (May 31 2012). At http://www.acceler8or.com/2012/05/when-i-called-charlie-stross-a-dirty-name-transhumanist/. Accessed 18 Dec 2014

Swierstra T, Rip A (2009) NEST-ethics: patterns of moral argumentation about new and emerging science and technology. In: Kaplan DM (ed) Readings in the philosophy of technology. Rowman & Littlefield, Lanham, pp 208–227

Swierstra T, van Est R, Boenink M (2009) Taking care of the symbolic order. How converging technologies challenge our concepts. NanoEthics 3:269–280

Swiestra T, Boenink M, van Est R (2009) Shifting boundaries. NanoEthics 3:213–216

Thacker E (2004) Biomedia. University of Minnesota Press, Minneapolis

The President's Council on Bioethics (2004) Beyond therapy: biotechnology and the pursuit of happiness. http://www.dana.org/news/cerebrum/detail.aspx?id=1286. Accessed 19 Dec 2014

Toumey C (2007) Privacy in the shadow of nanotechnology. NanoEthics 1:211–222

Toumey C (2008) The literature of promises. Nat Nanotechnol 3(4):180–181. http://www.nature.com.ezp.lib.unimelb.edu.au/nnano/journal/v3/n4/full/nnano.2008.74.html. Accessed 18 Dec 2014

Twist J (2004) Eco glass cleans itself with sun. BBC News Online. http://news.bbc.co.uk/2/hi/technology/3770353.stm. Accessed 17 Dec 2014

Venkatesan P (2010) Nanoselves: NIBIC and the culture of convergence. Bull Sci Technol Soc 30(2):119–129

Wang ZL (2010) What is nanotechnology? http://www.nanoscience.gatech.edu/zlwang/research/nano.html. Accessed 18 Dec 2014

Weckert J (2007) An approach to nanoethics. In: Hodge G, Bowman D, Ludlow K (eds) New global frontiers in regulation: the age of nanotechnology. Edward Elgar, Cheltenham, pp 49–66

Weckert J (2009) Nanoethics. In: Kyrre J, Olsen B, Andur Pedersen S, Hendricks VF (eds) A companion to the philosophy of technology. Wiley, Blackwell, pp 459–461

Wilsdon J (2004) The politics of small things: nanotechnology, risk, and uncertainty. IEEE Technol Soc Mag 23:16–21

Wolfe J (2005a) Top ten nanotech products. Forbes nanotech report, 1 December, 2005. http://www.forbes.com/2005/01/12/cz_jw_0112soapbox.html. Accessed 18 Dec 2014

Wolfe J (2005b) Safer and guilt-free nano foods. Available via Forbes. http://www.forbes.com/investmentnewsletters/2005/08/09/nanotechnology-kraft-hershey-cz_jw_0810soapbox_inl.html?partner=rss. Accessed 18 Dec 2014

Wood S, Geldart A, Jones R (2008) Crystallizing the nanotechnology debate. Technol Anal Strateg Manag 20(1):13–27

Ying HJ (2006) Managing the risks of nanotechnology development. Chin Sci Forum 5(September):110–113

Zhao Y (2007–2009) Director's note, Forefront. Annual report of IHEP Laboratory for Biomedical effects of Nanomaterials and Nanosafety, p 3

Chapter 3
Bioethics as an Approach to Nanoethics in China and the EU

'I've learnt from the GM debate', says Welland. 'It's easy to condemn a technology, but hard to fight back'. (Giles 2004)

Abstract This chapter looks at bioethics policy and regulation in China and the EU, examining societal values such as those of human rights, autonomy, and dignity in the West, and so-called 'Asian values' such as community, that underpin such regulatory approaches.

Keywords Confucianism • Western individualism • Bioethics • Dignity • Precautionary principle

The brief discussion of literature on nanotechnology issues in Chapter One offers some idea of what occupies the minds of nanoethicists. How else might we approach nanoethics? One argument is that nanoethics is an extension of bioethics debates. As Grunwald (2005) argues, 'many of the ethical questions raised by nanotechnology are already known […] The ethics of technology, bioethics, the ethics of medicine, [and] the theoretical philosophy of technology' already consider questions relevant to nanoethics. Nanoethics, given that nanomedicine and issues of risk to human health (toxicity) are the immediate issues with nanotechnology, is inevitably 'in some respects modelled on the development of bioethics' (Keiper 2007). While critics value the starting point offered by an already-documented field, they also appear to wish for something broader, though which remains difficult to define. Since a new and broader framework has yet to appear, one can argue that existing bioethics frameworks and narratives provide the most likely basis for nanopolicy and nanoethics development (Ryan 2004).

3.1 The GM Controversy

Before looking at bioethics, however, one might note that references to food ethics were constantly made in early discussions of nanotechnology issues (and indeed continue to be made), parallels being drawn with the introduction of GM food and

© Springer International Publishing Switzerland 2015

S. Dalton-Brown, *Nanotechnology and Ethical Governance in the European Union and China*, DOI 10.1007/978-3-319-18233-9_3

the subsequent public outcry in some EU countries.[1] It is therefore worth spending a little time on the controversy, and on the way it may show differing – or not – public attitudes in China and the EU, as well as public involvement in debates on the introduction of new technologies that might be potentially harmful while promising huge benefits.

Public response to GM food in the EU grew negatively with the publication of reports alleging that GM food was 'unsafe', that several GM crops had affected wildlife by reducing amounts of weed seed, and that GM crops offered little or no economic benefit (Mitchell 2003). Despite the fact that there is no evidence of toxicity relating to GM crops, as noted in a 2008 review published by the Royal Society of Medicine (Keys et al. 2008), the belief in the EU that this was a 'Frankenfood' was widely held in the 1990s. Following this public outcry over 'unsafe' food, the EU Council of Ministers placed a moratorium on GM food in 1998. This ban on genetically modified food imports is estimated to have cost the USA over 6 billion dollars in corn exports.

Studies in developing countries indicate that nanotechnology has the potential to advance agricultural productivity, particularly through more efficient fertilisers (Joseph and Morrison 2006). But in March 2008 Friends of the Earth called for 'a moratorium on the further commercial release of food products, food packaging, food contact materials and agrochemicals that contain manufactured nanomaterials until nanotechnology-specific regulation is introduced to protect the public, workers and the environment from their risks, and until the public is involved in decision making' (Miller 2006). A report from the Institute of Food Science and Technology in the UK argued that more safety data is required before nanoparticles can be included in food. The report pointed out that current legislation did not force companies to label food items containing nanoparticles (Joseph and Morrison 2006). In May 2011, the European Food Safety Authority (EFSA) published a guidance document for the risk assessment of nanomaterials in food and feed applications including food additives, enzymes, flavourings, food contact materials, novel foods, feed additives, and pesticides (Nanowerk 2011).

In China, while there has also been no definitive evidence showing GM food to be harmful to humans, the topic has been controversial for some time, escalating after the Ministry of Agriculture granted biosafety certificates to two pest-resistant GM rice varieties and a corn variety in November 2009 – a major step in promoting the research and planting of GM crops. The approval made China the first country in the world to give the nod to field trials of GM staple foods, but

> Officials, researchers and scientists are divided on whether and when GM food should be commercialized... Supporters and opponents have also been facing off over environmental

---

[1] The 2001 survey across the EU on GM food (averaged at 56 % against) indicated both an interesting variety of responses, from strong negative reaction in Britain and Belgium to lesser opposition in Spain and little in the Netherlands. Later surveys have shown a 'downward trend in support for GM food'. See Eurobarometer (2010).

safety and economic security issues. The Chinese government has put food security high on the agenda in its national development plans. (Xuequan 2010)

After granting these biosafety licenses, 120 Chinese academics signed a public petition in March 2010 asking the Ministry of Agriculture to withdraw the certificates (Jia et al. 2011).

A general Food Safety Law was only introduced in China in 2009, whereas one has been in place in the EU since 2002. The Chinese Law was 'pushed' by a food crisis relating to contaminated milk in 2008 and is regarded as 'still an ongoing process' complicated by 'considerable fragmentation of regulatory authorities', meaning that there 'is a considerable difference between theory and practice in the Chinese food safety system' (Polo 2011). There is also a lack of food safety watchdogs or other NGO bodies that could facilitate participation from the public, although it should be added that the Food Safety Law approval process was open to public scrutiny, and 'more than 11,000 comments were made to the law-making body' (Polo 2011). A *China Daily* 2010 survey concluded that more than 85 % of respondents were worried about the potential health hazards of GM food (Xinhua 2010). Hu and Chen's (2004) earlier survey of Beijing consumers found that consumer purchase intentions of GM vegetable oil were low, indicating a 'considerable scepticism toward GM products'.

Is public trust such a huge issue in China? Yes and no. A country with strong economic imperatives and an eye to global markets is always conscious that 'new' can be a selling point but that 'untested and potentially unsafe' is unacceptable to risk-averse customers. However, there is less of a tradition of public participation in policymaking. Thus it is difficult to see what impact the public protests may have had. The Chinese government is still investigating whether the town of Jiangkou, located in Central China's Hunan province, was allegedly used as an experimental site in 2008, with 'dozens of children believed to have been fed GM modified rice as part of a nutrition research program led by a professor from Tufts University in the US' (Xinhua 2012). There have been small-scale protests, as when around 40 people, demonstrating outside the Ministry of Agriculture in September 2011, presented a letter asking the ministry to stop advocating staples such as GM grain in China. About 80 people had signed the letter, including experts and some former government officials (Shuang 2011).

The two main lessons of the GM debate appear to be that there are issues about public involvement in the process – more participation might have encouraged less public paranoia – and that new technologies can become 'plausibly linked to catastrophic scenarios of new and unpredictable technologies gone awry' (Lin 2006).[2] The GM debate created a considerable trust deficit among the global public, one that may paralyse nanodevelopment again (van Calster 2008).

---

[2] See also Ebbesen (2007) and Moor and Weckert (2004).

3.2 Back to the Bioethics Debate

The most common approach to nanoethics thus far in the West has been that of bio-ethics, which can be defined as 'the systemic study of human conduct in the area of life science and health care, insofar as this conduct is examined in the light of moral values and principles' (Reich 1995).

Bioethics has consistently posed some of those speculative 'big questions' about how an ethical relationship between technological advances and human beings might be defined. When bioethics emerged in the USA as a reaction to the biotech-nological and biomedical advances of the 1960s, its questions were very similar to those currently identified in the context of nanotechnology, namely,

> How were human beings wisely to confront the moral puzzles, perplexities, and challenges posed by the confluence of the great scientific and cultural changes?…Who should have control over the newly emergent technologies?… How could individuals be assisted in tak-ing advantage of the new medical possibilities or, if need be, protected from being harmed by them? How could the fruits of the medical advances be most fairly distributed? What kind of character or human virtues would be most conducive to a wise use of the new tech-nologies? What kind of institutions, or laws, or regulations would be needed to manage the coming changes in a moral fashion? (Callahan 2002)

André Hellegers and Van Rensselaer Potter were the first to use the term 'bioeth-ics' in the 1970s to designate a focused academic area of inquiry, noting that the overriding question was that of medical technology's effects on society (Reich 1994). The Kennedy Institute established by Hellegers (and its counterpart, the Hastings Institute) appears to approach bioethics as more than a study of ethical issues in medicine. They also include issues about public health, population con-cerns, genetics, environmental health, reproductive practices and technologies, ani-mal health, and animal welfare to name just a few.[3] In this sense, bioethics offers a good starting point for a discussion of how different cultures may approach the broad issues offered by a potentially far-reaching new technology such as nanotechnology.

We shall therefore use bioethics as a starting point to see what values inform the bioethics debate, looking for any similarities that might exist in China and the EU.

The following discussion will cover four topics in both regions: (1) social values, (2) bioethics guidelines, (3) guiding principles, and (4) public attitudes to science and technology.

Given that the point of this study is to argue that cultural differences can be to a degree overcome in a public decision-making process, the following discussion sets the parameters of difference and similarity. It may seem, given the accepted (if cli-chéd) understanding of the differences between the East and the West, an impossi-ble task – how can one see any real convergence between a highly individualistic

---

[3] The Hastings Centre, established in 1969 (Hellegers established the Kennedy Institute of Ethics in 1971), saw its mission as 'to address fundamental ethical issues in the areas of health, medicine, and the environment as they affect individuals, communities, and societies'. See http://www.the-hastingscenter.org/About/Default.aspx

Western culture based on a strong human rights discourse and a Chinese culture in which social harmony is key and with greatly differing political structures from those in Western democracies? Nevertheless, this study will attempt to draw out the shades of grey between the black-and-white differences usually accepted in discussions about the East and West.

3.2.1   Social Values and Bioethics: Informing the Nanodebate in the EU

> Do we today have an available bioethics? Yes, we do, a bad one: what the Germans call Bindestrich-Ethik, or 'hyphen-ethics', where what gets lost in the hyphenation is ethics as such. The problem is not that a universal ethics is being dissolved into a multitude of specialised ones (bioethics, business ethics, medical ethics and so on) but that particular scientific breakthroughs are immediately set against humanist 'values', leading to complaints that biogenetics, for example, threatens our sense of dignity and autonomy. (Žižek 2003)

We begin with an obvious rider – to refer to the EU as having 'one approach' to bioethics is of course far too general. The EU, created as a political and economic union of now 28 member states, is usually seen as an economic entity rather than a cultural one. That said, given the current issues in the Eurozone and the threat of withdrawal, those economic ties might be looser than was first thought. In a more general sense, it is fallacious to assume that cultures themselves do not contain wide heterogeneity, and so it may be no more difficult for a 'US bioethicist to discuss ethics with an Asian colleague than for a US Mormon to discuss them with a Jewish neighbour' (Solomon 2006).[4] And it is equally fallacious to state that cultures are static, for they contain 'a diversity of specific life-forms, each with its own peculiar laws of evolution' or 'ongoing conversations' (Gbadegesin 2007; Harbour 1999). Discussing the values of EU culture and society is therefore an inevitably partial endeavour; yet at the same time there are agreements commonly seen as 'Western' or 'European' that provide a basis for comparison with other global regions.

3.2.1.1   EU Social Values

What agreement might there be on European values? President Barroso (2004–) summarised them at the 10th South East Europe Cooperation Process Meeting in Zagreb in 2007:

> This year we celebrate the 50th anniversary of the European Union. We are proud of this achievement, which has brought peace, prosperity and solidarity to a continent wracked by

---

[4] Solomon sees three aspects to this problem: the temporal export problem; the problem of exporting ethical insights among persons in the same culture, i.e. the local export problem; and the problem of translating ethical insights within the ethical viewpoint of a single person, i.e. the personal export problem.

war [...] we reaffirmed our shared values, like freedom, democracy, the rule of law, tolerance and mutual respect. .... Those values formed the very foundations of the European
Union. They remain at the core of our activities today.

Barroso's list of values suggest the oft-declared basic principle of European and
Western legislation, that of individual rights. Dignity, freedom, equality, solidarity,
citizens' rights, and justice are the six main ideas of the 2000 European Union
Charter of Fundamental Rights. Dignity, freedom, democracy, equality, the rule of
law, and respect for human rights are the core values that were set out by the EU at
the beginning of the Treaty of Lisbon (2009), which strengthened the role of the
European parliament and increased European co-decision-making on policy. Given
these statements, it is unsurprising that approaches to bioethics in the EU tend to
'emphasise individualistic ways of thinking' (Scott 2009).

3.2.1.2   EU Bioethics Guidelines

These values are replicated in four international (thus not specifically European)
biolaw Conventions and Declarations, adopted by nearly all member states of
the EU:

1. The Convention on Human Rights and Biomedicine, or 'Oviedo Convention'
   (1997)
2. The Universal Declaration on the Human Genome and Human Rights (1997)
3. The International Declaration on Human Genetic Data (2003)
4. and the Universal Declaration on Bioethics and Human Rights (2005) (Andorno
   2007)

The Oviedo convention, as its full title (Convention for the Protection of Human
Rights and Dignity of the Human Being with regard to the Application of Biology
and Medicine: Convention on Human Rights and Biomedicine) suggests, stresses
human rights and dignity as overarching principles of EU bioethics (Lenoir and
Mathieu 1998).

The classic Western liberal notion of human rights emphasises absolute individual political and civil rights, while most non-Western traditions place greater
emphasis on the community basis of rights and duties and on economic and social
rights. This may be the key difference between the EU and China in approach to
lawmaking, including biolaw.

Yet one should beware the cliché of an apparent antagonism between Western
human rights/individualism and a more communal, socially determined form of
existence in the East, i.e. the so-called Eastern communitarianism. As Bielefeldt
(1995) argues, 'what is at stake in human rights is not an abstract individualism, but
rather the principle of equal freedom, which, as a critical demand, always affects
individuals and communities simultaneously. Thus he argues that

...although human rights clearly enlarge the scope of individual freedom, they are by no
means merely individualistic. They are not meant to lead to an 'atomistic society' devoid of

communitarian solidarity. Against the widespread confusion of human rights and Western individualism, human rights always imply a social dimension because human freedom can unfold only in relation to fellow persons.

We can interpret the notion of 'human rights' rather more broadly than as 'democratic freedom' or 'democratic protection for the vulnerable'. Traer (1991) argues that one might 'look beyond the institutions Westerners equate with human rights' to those 'cultural forms' that set forth 'those political, social, and economic rights that contribute to the dignity of the individual person'. Bielefeldt (1995) similarly argues for 'pluralism and difference' as applicable to the concept of human rights (while acknowledging that some cultural practices are precluded under any form of human rights – an example being that of slavery).

Perhaps instead of institutions, we should follow the example of the bioethics Declarations, in which the application of principles is given precedence over concrete rules.

Of the guiding principles of autonomy and dignity, we shall, perhaps controversially, given that autonomy is often claimed as the chief principle in bioethics, examine the second – dignity.

3.2.1.3 Guiding Principles: Dignity and Other Principles

Before looking primarily at dignity, we should consider some of the other principles that have been noted. 'Principlism' was first formalised as a moral decision-making approach in the USA, by the National Commission for the Protection of Human Subjects of Biomedical and Behavioral Research in the Belmont Report on April 18, 1979. The report resulted in a statement of four basic ethical principles. Beauchamp and Childress (2001) have promoted these four guiding ideas: not inflicting harm intentionally (nonmaleficence), assisting the individual to make meaningful choices (autonomy), acting so as to contribute to the welfare of others (beneficence), and offering fair, equitable, and appropriate treatment (justice).

There have been similar 'restatements' of these principles such as the BIOMED II project, Basic Ethical Principles in European Bioethics and Biolaw (1995–1998), a cooperation across most EU countries that identified the principles of respect for autonomy, dignity, integrity, and vulnerability as four important ideas or values for European bioethics and biolaw. They thus serve as reflective guidelines and important values in European culture (Rendtorff 2003).

Although Gert et al. (1997) argue that the principles are often mere 'checklists', certainly not all clear action guides, they do provide a practical approach for many of the health ethics issues faced by practitioners ('checklists' are not at all a bad thing).

Why dignity, then, as the centre of our discussion, rather than autonomy?[5]

---

[5] Macklin (2003), however, sees dignity as nothing other than autonomy.

Instead of arguing for the priority of autonomy over dignity, this section will look at dignity as a major principle of bioethics and one that might allow a point of similarity with

China. Whereas China does not value autonomy as highly, nor does it follow the human rights discourse enshrined in the EU regulation, dignity is a term relevant to Chinese culture, as it is in the West. This might offer some way forward in terms of global ethics, moving us beyond the individual rights/autonomy versus state power over the individual dichotomy only too apparent from any political analysis of the two regions.

Andorno (2007) notes that the principle of recognising human dignity is supported in the above-mentioned Declarations by a stress on the primacy of the human being over the sole interest of science or society. In addition, the European Convention on Human Rights and Biomedicine states that parties to the Convention shall 'protect the dignity and identity of all human beings' and shall guarantee everyone, without discrimination, respect for their integrity and other rights and fundamental freedoms with regard to the application of biology and medicine (Council of Europe 1997). Human dignity, as a central 'value' or principle might seem to give EU bioethics a distinctive individualistic flavour, one that can be contrasted to the so-called Eastern approach in which societal good is prioritised over that of the individual.

Yet *is* dignity solely individualistic? Kantians would argue that the term relates to the group dignity of 'rational beings'. Dignity is a confused term in contemporary bioethics. It has been used by American bioethicists to refer to autonomy and by Catholic bioethicists to refer to the sanctity of life, to name just two contradictory meanings. Macklin's 2003 'Dignity is a Useless Concept' suggests a practical approach would be preferable, arguing that instead of a nebulous concept (or what she terms a 'slogan' or 'restatement of other vague concepts'), the recognition that no one has the right to impinge on the life, body, or freedom of the other works better for practitioners. In this context, dignity becomes synonymous with autonomy (Pinker 2008).[6]

Dignity can refer to 'both the intrinsic value of the individual and the inter-subjective value of every human being in its encounter with the other' (Kemp and Rendtorff 2000). The second phrase suggests the social nature of dignity alongside its more individualistic meaning.

I follow Schroeder's (2012) approach, which separates dignity and human rights.[7] Schroeder's work on dignity suggests that two categories can be distinguished – inviolable and aspirational dignity. Her analysis of Kantian dignity, which states that it is the inviolable property of all rational beings to be treated as an end, not merely as a means, differentiates it from aspirational dignity, in terms of which there are certain expectations of an individual determined by society and culture (Schroeder 2010). For example, in the euthanasia debate,

---

[6] Steven Pinker's development of Macklin's idea argues that it has been hijacked by conservatives.

[7] For Schroeder's compelling arguments for such separation, see Schroeder D (2012).

Table 3.1   Distinct meanings of dignity (Schroeder 2012)[a]

Individual dignity	Traditional Catholic dignity	Dignity is an inviolable property invested by God in all human beings, which makes each life sacred
	Kantian dignity	Dignity is an inviolable property invested in all rational beings due to their capacity for moral self-legislation. As dignity holders, rational beings have the right to exact always respect for their sense of purpose and self-worth
Aspirational dignity	Aristocratic dignity	Dignity is the quality of a human being who has been invested with superior rank and position and acts accordingly
	Comportment dignity	Dignity is the outwardly displayed quality of a human being who acts in accordance with society's expectations of well-mannered demeanour and bearing
	Meritorious dignity	Dignity is a virtue, which subsumes the four cardinal virtues[b] and one's sense of self-worth

[a]My thanks to Professor Doris Schroeder for sending me this table
[b]The four cardinal virtues are: prudence, or appropriate action; justice; restraint (self-control, moderation); and courage (or fortitude)

one view is that everyone has the right to aspirational dignity and thus to die if that dignity is traduced. The opposing view is that the individual must continue to live due to his/her inviolable dignity. Such views are not only contradictory; they show one of the many possible disagreements about the meaning of dignity. The following table summarises the different meanings as identified by Schroeder (Table 3.1).

Looking at the above distinctions, one could understand aspirational dignity as individualistic. For instance, individuals strive to attain the four cardinal virtues and a sense of self-worth. However, inviolable dignity works more at the group level or even species level, as shown in Kant's definition of all *rational* beings having dignity or the Catholic Church's belief that all *human* beings have dignity.

In a similar vein, Feldman (1999) argues that dignity has three spheres of operation, consisting of the dignity of the species, of a group, and of the individual. Human dignity can be universal, or subjective, or socially relative.

In conclusion:

1. Dignity is a term that can refer both to the individual's rights and to his/her social interactions.
2. Whereas one might argue for the dignity of the *individual* as more important in a Western values system and the dignity of the *individual in social terms* as more important in the East, a definition of human dignity can include both.
3. Dignity is a concept that can be discussed in terms of methodology and virtuous agency, rather than as simply value based. In other words, dignity can be expressed not as autonomy but as agency, as the second half of this study will discuss.

3.2.1.4   EU Public Attitudes to S&T, and the Precautionary Principle

Given that social values can be significant drivers for a citizen's attitudes towards science, the 2001, 2005, 2008, and 2010 EU (*Eurobarometer*) surveys of S&T attitudes might be expected to emphasise these core values. How interested are citizens in S&T, though? A *Eurobarometer* survey in 2002 on European public attitudes to biotech concluded that on the whole, the European population is largely inattentive to biotech advances but is globally positive about biotech (Gaskell et al. 2003).[8] Other surveys, however, have suggested ambivalence (European Commission 2010).

A 2010 survey noted that Europeans 'express interest in new scientific discoveries and technological developments where 30 % are very interested and 49 % are moderately interested' (Eurobarometer 2010). The key findings of the 2010 survey confirmed that Europeans appeared moderately interested and moderately well informed on new technologies, although the latter tends to vary according to which technology – the 2002 *Eurobarometer* report noted a 'surprisingly low' level of understanding of some basic biotechnology issues, and nanotechnology continues to be less well understood than many other relatively new sciences (Gaskell et al. 2003). There is also, of course, regional variation; in the 2010 survey, six countries in the EU showed quite low levels of interest in technology – Poland, Portugal, Romania, Lithuania, Turkey, and Bulgaria. There was some correlation here with countries that regard themselves as poorly informed on science and technology, such as Bulgaria, Romania, Portugal, and Turkey (others being France, Luxembourg, Austria, Slovakia, and Spain) (Eurobarometer 2010).

The more pertinent, though very broad, question is perhaps whether the public believe that science is valuable and ethical. Some states have a cultural distrust of scientists 'playing God' which might influence that view, and in this context, one might note that 2 out of 5 European respondents admitted to being superstitious (with the highest rates being recorded in Italy, the Czech Republic, Latvia, and Slovakia) (Schummer 2006).

In the 2010 survey, general negativity about the effects of science manifested in scepticism towards scientists who 'cannot be trusted to tell the truth about controversial scientific and technological issues because they depend more and more on money from industry', while certain countries appeared to have become more cynical – Germany being 30 % less positive on S&T than in previous surveys, for example. Six out of ten Europeans felt that S&T can sometimes damage people's moral sense, and 1 in 2 thought that some applications of S&T can threaten human rights. Europeans also felt that S&T could be used by terrorists in the future. When asked which area of research should be prioritised by researchers in the European Union, 40 % of respondents mentioned health issues, with energy issues at 21 % and environmental issues at 18 %.

Only slightly more than half of Europeans surveyed agreed that new inventions would always be found to counteract any harmful use of science. Half felt that if a

---

[8] The survey added that 'the European public employs a sort of "risk rhetoric" that stays largely on the declarative or ritual plane and exerts little real influence on the perceived "usefulness" of applications'.

new technology poses a risk that is not yet fully understood, then its development should be stopped even if benefits are expected.[9] Thus European technology policy has therefore focused on the principle of protecting individual human rights, chiefly through 'precaution':

> Following an assessment of available scientific information, where there is reasonable concern for the possibility of adverse effects but scientific uncertainty persists, measures based on the precautionary principle may be adopted, pending further scientific information, for a more comprehensive risk assessment, without having to wait until the further reality and seriousness of those adverse effects become fully apparent. (Guston 2010)

The precautionary principle suggests some counterbalance to economic imperatives, as shown when looking at environmental threats:

> In order to protect the environment, the precautionary approach should be widely applied by states according to their capabilities where there are threats of serious or irreversible damage, lack of full scientific certainty shall not be used as a reason for postponing cost-effective measures to prevent environmental degradation. (Petroni 2006)

The precautionary principle underscores the EU's focus on individual rights, as does the general European concern for increasing the level of public involvement in S&T decision-making, with leaders in this field being Holland, the UK, France, and Germany. However, the 2010 *Eurobarometer* survey noted that 91 % of respondents either never or hardly ever attended public meetings or debates yet felt that governments should do more to encourage young people and women to be involved with science. It seems that the level of public disinclination to be actively involved in S&T debates (instead of being 'passively' surveyed) is still high.

In conclusion, although discussing a homogenous EU approach to bioethics can only lead to partial statements, one might predicate a strong narrative based on protecting the individual's rights and dignity. This is perhaps reflected in the degree of public scepticism towards S&T and the prevalence of the precautionary principle, which allegedly places risk to individual human health and the environment higher than economic imperatives.[10]

3.2.2 Social Values and Bioethics: Informing the Nanodebate in China

> Ought bioethics in East Asia to use the same approaches (assumptions, principles, theories, styles, methods, concepts) as bioethics developed in the West, or ought it to reflect a specifically East Asian approach to the subject? (Fan 1997)

---

[9] And finally, a majority of Europeans believed that collaboration between the EU and other countries is important to make the EU a global player, 7 out of 10 believing that joint research collaboration with the USA is important, 64 % thinking that links with poorer countries should be strengthened, and 61 % of respondents advocating links in particular with China and India.

[10] Weckert and Moor (2006) offer a useful article defending the precautionary principle's use as a coherent tool in nanotechnology assessment.

The question given in this quotation suggests that we must again begin with a rider: Asia is composed of 3.4 billion people of diverse ethnic, linguistic, and religious composition (Confucian, Buddhist, Islamic, Hindu) and thus it is unlikely that Asia subscribes to a single set of beliefs, completely different from those held by a billion people in Europe and America. Thailand, for example, has a controversial history in biotechnology, ranging from issues of morality and environmental concerns, through to issues of intellectual property such as 'biopiracy' and compulsory licensing (Changthavorn 2003).[11] Other factors are those of urbanism and education with Pollard (2007) raising the issue of country attitudes given that 'science and technology are formally accepted ideas among only sophisticated Thais in metropolitan areas'.

It should be noted that Asian values are often 'unified' by being conflated with Confucianism; Bell (2006) insists that even differences across Asian cultures are rooted in Chinese political 'traditions' such as Confucianism.

3.2.2.1 Asian Social Values

Leaving this aside, one can still offer some general comments on Asian social values. According to various surveys, in which Asian respondents defined the top six societal values, in their view, two, unsurprisingly, were collective values such as an orderly society and societal harmony (the others were accountability of public officials, openness to new ideas, freedom of expression, and respect for authority) (Yu 2003). The Asian values movement (as promoted by Lee Kuan Yew, former Prime Minister of Singapore, and Dr Mahathir bin Mohamad, former Prime Minister of Malaysia) argues strongly for communitarian values. Lee has attributed Singapore's speedy economic achievements to such Asian values as 'strong family ties and responsibility for the extended family', hard work, thriftiness, and individual discipline for family benefit. Mahathir has denounced Western individualism for having led to 'the breakdown of established institutions and diminished respect for marriage, family values, elders, and important customs, conventions, and traditions' (Barr 2002).[12]

Critics of the concept of Asian values (some 'would claim that the notion... has served as a pretext for soft authoritarianism') (Kim 2012) in fact repeat some of these values, such as the importance of family, communitarianism, work ethic, and so forth (Kim 2012).[13] Bell's (2006) claim that there 'are no distinctly Asian values' does not seem commonly supported.

Thus the general assumption about 'Asian values' is that collective harmony is the core, as opposed to the Western emphasis on the individual and his/her rights, autonomy, and dignity. Wang (1997) notes a Confucian emphasis on 'care for others', in terms of Confucian *ren* or *jen* ethics, meaning less focus is placed on the

---

[11] See also Kachonpadungkitti and Macer (2004).

[12] Yew and Mahathir are both cited in Barr (2002). See also Widdows (2011).

[13] See also Thompson (2001) and Emerson (1995).

independent person. Certainly family is important in many Asian nations, and this may lead to various inequities from a Western perspective; for example, the Singapore government advocates unequal distribution of medical benefits to male and female employees. Such a policy is unjustified from the perspective of gender rights, but it can be justified if the father is regarded as having special status, to which the makers of social policy owe a special responsibility. The Prime Minister of Singapore, Goh Chok Tong, has argued that the government would like 'to channel rights, benefits, and privileges through the (male) head of the family, so that he can enforce the obligations and responsibilities of family members' (Steiner and Alston 2000).[14]

Perhaps proponents of Asian values simply wish to ensure a placid, cooperative, and hardworking population; however, the movement does offer a means for deflecting accusations of human rights abuse. Democracy may always be interpreted rather differently in the East compared with the West; a paper analysing the *East Asia Barometer (EAB)* survey conducted in South Korea in February 2003, looking at whether Western-style liberal democracy is compatible with Confucianism, concluded that

> …contemporary Korean political culture still manifests the Confucian legacy of hierarchical collectivism and benevolent paternalism, the Confucian ideal of family still remains a model of governance in the eyes of many ordinary Koreans. Yet those attached to Asian values still desire democracy… however, their view of democracy might be one of good communitarian governance. (Park and Shin 2004)

No wonder that the 1993 Bangkok Declaration on Human Rights affirms the universality of human rights but pleads for greater context sensitivity in their promotion (Sullivan 2007).

The *AsiaBarometer* 2003, 2004, and 2005 surveys (each of which was carried out across 6–12 countries ranging from Japan to Laos to Mongolia to Uzbekistan) covered a variety of topics.[15] One commentary on the survey data, the collection *How East Asians View Democracy*, argues that democracy is in fact seen as an important value in Asian societies – but when placed against economic development, it becomes rather less so (Chu et al. 2009). The need to provide for one's family is the chief motivation for individuals. The same commentary also notes that the doctrine of personal interests coming in second to those of the community is becoming less popular; responses to the statement, 'Even if parents' demands are unreasonable, children should still do what they ask,' which used to be an Asian value, indicated that such family obedience was less popular today (Nathan et al. 2008). A 2005 study that asked respondents to state what was the 'most important thing' in their lives found that the most frequent answer was 'family and children'

---

[14] Such a policy is regarded as having the effect of strengthening the family.

[15] Such as infrastructure development, economic conditions, life values, customs, rules, identity, political consciousness, and health (the 2005 survey did not include health conditions). The 2006, 2007, and 2008 surveys added the following topics: quality of life, governance, democratic consolidation/regression, social virtues, happiness, international alignments, new middle class, religiosity, mass media, and globalization.

in Japan and South Korea, but in Taiwan, Beijing, Shanghai, and Hong Kong it was 'life, health, myself' (Minato 2007).

China's Maoist communitarianism is still the dominant paradigm since the formation of the People's Republic of China in 1949, though there has since evolved a form of hybrid

> socio-political philosophy... based on two thousand years of power-centralized, autocratic monarchy – one that has lacked any rights-oriented, individualistic, liberal democratic tradition... The historicism and social holism of this system, interwoven with traditional ideas, puts the greatest emphasis on nation, society and country rather than on individuals. (Qiu 1992)

The notion of community is actually broader than that of family, and means 'harmony'. In traditional Chinese society,

> ...there is less emphasis on individual rights, self-expression, and self-determination. In the community, qualities such as harmony, function, and responsibility are stressed more than individual rights, and familial relationships assume primary importance. (Ip et al. 1998)

3.2.2.2   Chinese Bioethics Guidelines

Bioethics became a discipline in China following the first Chinese euthanasia trial in 1986, the subsequent debate about a person's right to life, and the publication of Qiu's *Bioethics* in 1987.[16] However the label of 'Chinese bioethics' must be used with caution, as bioethics itself, as one critic has claimed, is a Western idea not simply because it originated in the USA or has its roots in the West but because of the way it is theorised, structured, formulated, and practiced, as a 'normativity of whiteness' (Myser 2003).[17]

In China, bioethics was not initially regarded as a priority:

> The fact that Chinese scientists don't share the ethical concerns of their Western counterparts is confirmed by a 1993 survey of 255 Chinese geneticists: An overwhelming majority said that public health and the "quality" of the population should be improved through practices that would be rejected in the West as eugenics.... 'Chinese culture is quite different'," said the Chinese scientist who performed the survey. 'Things are focused on the good of society, not the good of the individual. It would shock people in the West, but my survey reflects cultural common sense'. (The New Atlantis 2003)

However, bioethics has recently become more 'respectable' in China, as Guo (2011) states:

> The importance of international collaboration in science has led policymakers to recognize the value of 'ethics' for purposes of global recognition. The government's awareness of subject (sic) has improved the respectability of bioethics as an academic discipline.

In 1998, the Chinese Ministry of Health (MOH) issued its *Interim Regulations* or *Ethical Review of Biomedical Research Involving Human Subjects*, defining proto-

---

[16] A general introduction to ethics in relation to genetics, euthanasia, organ transplantation, etc.

[17] See also Arekapudi and Wynia (2003).

cols for an ethical review of human biomedical research in China. Research institutions would establish ethics committees to safeguard procedures in human subject research, while all large certified hospitals would establish medical ethics committees. In addition, the MOH established a national ethics committee and provincial ethics committees. China has regulations in place on organ transplantation, scientific misconduct, and AIDS treatment. Ethics committees are established in most large Chinese hospitals and thus it has been argued that 'medical professionals and the highly educated populations in the PRC are mostly influenced by, and accept, Western bioethics' (Wang 2011).[18] Although China now has regulatory structures similar to those of the West, Hennig (2006) claims that they differ in that they 'are not enforceable by law'. However, Qiu (2011) suggests that 'after years of debate, China has reached a consensus that it is indeed necessary and desirable to regulate biomedical research and biotechnology for the purpose of protecting human subjects'. Thus the 2007 *Regulation for Ethical Review of Biomedical Research Involving Human Subjects* requires that each institution conducting biomedical research must establish an Institutional Research Ethics Committee (IREC), providing a framework that will protect human subjects. Yet Qiu (2008) notes that 'a number of challenges have arisen during the course of implementing the Regulation', due to resistance to ethics and malfunctioning ethics committees.

There are other instances in which the Chinese approach to bioethics differs from that of the West. In terms of protecting the human subject, China has guidelines on Human Assisted Reproductive Technologies (MOH, July 2003) that prohibit human reproductive cloning. By contrast, the country does allow the creation of human embryos for research and therapeutic purposes like embryonic stem cell research, leading to the argument that the Chinese cultural environment has far fewer moral obstacles to the use of human embryos in research than many other nations have (Qiu 2004).[19] Klein (2010) sums it up as follows:

> In China's race to the top, the... central government therefore targets specific selected areas with the potential to compete successfully internationally. Human embryonic stem cell research (HESCR) is one such area, because cautious stem cell policies and the ongoing controversy in large parts of the developed world offer unprecedented research and commercialization opportunities for China... due to Confucian moral philosophy Chinese have no moral qualms in China about using human embryos in stem cell research. As personhood can only be acquired through social practice, according to Confucian teachings, human value evolves out from an individual's social relations with society, the family or other groups, but it cannot be acquired before birth.

Hence the view that the EU, given its increasing trade and cooperation with China, needs to 'persuade China to adopt the precautionary principle as part of its environmental and trade policy', in order to facilitate the 'harmonization and implementation of EU and Chinese regulations and standards' (The National Foreign Trade Council 2003).

---

[18] See also Li (2008).
[19] See also Yang (2004) and Wang (2003).

3.2.2.3 Communitarianism and Other Principles

Chinese culture is seen as one of obligation, not one based on individual rights. This notion has become so widespread that it is rarely challenged. Yet the difference may be one in *degree* rather than substance – the so-called Asian shift away from individual rights does not necessarily *negate* individual rights but rather recognises the limits of the individual rights-based approach for people who have a relational theory of personhood. As Wertz et al. (2003) argue,

> Although Western Ethics is based on rights and principles and Asian ethics is based on caring and relationships, often the practical outcomes of the two approaches are similar.... Western principles of non-maleficence, beneficence, and justice are implicit in the Confucian ideal of humanness. The difference between Asian and Western ethics lies principally in the amount of credence given to the autonomy, privacy, and rights of atomized individuals.

Lee (1992) argues that Chinese society *needs* Western individualism to counter 'excessive emphasis on the collectivist conception of the common good', in terms of which 'people's assertions of basic rights and freedom have been neglected'. Thus, although the notion of community is not quite as inflexible as may be thought, it is still prioritised over European individual rights. Chattopadhyay (2008) argues that there are two particular divergences of approach:

> Consider autonomy, the dominant principle of bioethics in the United States... It is difficult for some Western bioethicists to realize that the principle of individual autonomy – even when tweaked to fit in non-United States settings by relocating autonomy in the family, or the clan, or an elder – is an assault on the tradition and values of non-Western societies who believe in the matrix of relationships in dynamic equilibrium of the cosmos. Or, consider the language commonly used in bioethics such as 'end-of-life'. This term, so widely accepted in the West is heard and apprehended much differently by a traditional Hindu who believes in 'life after death'.[20]

Fan (2006) draws a distinction between Western and Eastern concepts of autonomy, in that he first requires that the patient, as long as being competent, has the final authority to make clinical decisions for himself, whereas in Asia

> the family's decision should be made for the best interests of the patient in accordance with the objective conception of the good life adopted by the local cultural group. And it is the value of harmonious dependence between family members, rather than individual differentiation and independence, that this principle upholds.[21]

Thus Kaelin (2009) contrasts the UNESCO Universal Declaration on Bioethics and Human Rights, with its commitment towards universally shared ethical values, with the reality of the Filipino notion of autonomy, which does not focus on 'individual consent to health care, individual confidentiality, or individually articulated concerns with beneficence, caring or truth-telling'.[22] In terms of medical ethics, 'the family, more than the individual, is often considered as one basic unit in the two

---

[20] See also Stonington and Ratanakul (2006).

[21] See also Fan (1998).

[22] See also Alora and Lumitao (2001).

aspects of doctor-patient relationships' (Tai and Chung 2001). As Fan (2006) states: 'the family, having a responsibility, must shoulder the fiduciary obligation of care for the patient, including taking care of the burdens of communicating with the physician and making medical decisions and signing a consent form for the patient'.

The Asian view of autonomy is much more relational than in the West. It is unsurprising therefore that Asian bioethics is informed by recognition of the inter-dependence of all forms of life on earth, a 'holistic harmony' – 'community' in a very broad sense (Sakamoto 1999). In China, for example, *tao*, or harmony with the universe, is seen as the context for social harmony:

> Confucius recognized that not everyone was willing to cultivate a high degree of moral virtue, and that not everyone who tried would succeed. In practice, Confucius seemed to believe that only a minority would succeed in attaining moral superiority, and that he should concentrate his effort on enabling such a moral elite to become good leaders of the rest... (yet) in Confucian thinking, the dignity of the individual comes not from the capacity to act independently of other members of society but from the capacity to be a part of an interde-pendent whole. (Madsen 2007)

The reference to dignity is important here, given the previous contention that the EU bioethics declarations emphasise dignity. De Bary (1998) suggests that when discussing the Chinese individual and his/her dignity, we use the term *personalism* rather than that of individualism, as it suggests less of the Western context of liberal values and more of the worth and dignity of the person as 'self shaped and formed in the context of a given cultural tradition, its own social community, and its natural environment to reach full personhood'. Rather than placing collective over individ-ual, the Confucian way is to seek a *balanced* relation between self and society.[23]

Dignity in the Chinese context means various things. Rather than autonomy and individual rights, it means agency, self-cultivation, and compassion. In fact, this description aligns well with what Schroeder describes as meritorious dignity, which requires individual agents to strive for the cultivation of virtues, as well as more universal 'humanness', for 'the belief in human dignity presupposes an irreducible worth attached to every person insofar as s/he is a human being' (Zhang 2000).

It has been argued that dignity has always been and is becoming more important (at least 'officially'):

> The Chinese government places great importance on respecting human dignity. In April 2006, President Hu Jintao said... that 'Chinese civilization has always given prominence to the people and respect for people's dignity and value'... Premier Wen Jiabao pointed out that 'everything we do, we do to ensure that the people live a happier life with more dignity and to make our society fairer and more harmonious'. (China Daily 2011)[24]

In her analysis of Chinese historical texts, Fox Brindley (2010) sees agency as individual potential balanced by external checking agencies, holistically combined.

---

[23] Liu (1993) argues that the term 'geren zhuyi' (individualism), introduced into China in the early twentieth century, was hijacked by the state as a useful metaphor for a negative West, for the state 'has a political stake in presenting the idea of individualism to its people as un-Chinese'.

[24] A report of the speech by Luo Haocai, President of the China Society for Human Rights Studies, at the opening ceremony of the 4th Beijing Forum on Human Rights on September 22, 2011.

Although we cannot necessarily prove such a view to be reflected in contemporary forms, it suggests a perspective of agency and individuality that is more complex than merely 'bowing to authority'. While the Chinese view does differ from the Western-centric one of the individual as possessing sovereign rights and full autonomy, the Chinese holistic individual can 'achieve salvation for himself or herself from within a much larger web of social and cosmic interconnectedness and agency'; rather than 'rights-bearing' individuals, they are 'agency-fulfilling' individuals (Fox Brindley 2010).[25]

'Self-inspired' agency is linked to self-cultivation or to what Angel (2002) calls 'self-mastery'. As Ivanhoe (2000) argues, the Chinese 'enduring concern with the issue of moral self-cultivation' can mean either the Confucian model of acquired virtue or that of the Mengzian developmental model. Ivanhoe's work is interesting in that it underlines the notion that although virtue is to be cultivated for the greater good of the community, it is first and foremost a focus on the self. In his view, Confucian virtue acquisition rests on the creation of a separate or individual identity. Thus 'the conventional wisdom that the Chinese are group-oriented is, paradoxically, matched by the equally conventional view that the Chinese are individualists', the latter deriving from the neo-Confucian doctrine of individual perfectibility (Pye 1996). Opposing this view, Schwartz (1985) notes that although Confucianism argues for self-realisation, it does not stress liberty or individual rights. However this perpetuates a very Western way of thinking about Chinese individualism, trying to shoehorn it into Western concepts.

Ivanhoe (2000) suggests that the focus on self-cultivation is both an individual and a universal process. Discussing the philosopher Dai Zhen's version of Confucianism, he notes Dai's argument that 'one needs to pass one's spontaneous reactions to things and events through the sieve of universalisability in order to filter out "mere opinions" and arrive at the "unchanging standard" of moral truth'. The Confucian idea of acquiring a social, humble form through the rules of relationships (or social etiquette) sounds old fashioned yet contains a modern idea that one has a 'social' self that directs individual virtues towards consensus. I will come back to this idea when I deal with the issue of the type(s) of agency required by the Habermasian agent. Part of the drive towards consensus comes from the notion of compassion, prominent in Asian professional ethics (Sass and Xiaomei 2011) but, more broadly, from the Chinese notion of *ren* or *jen*, a compound term meaning 'humanness' and encompassing benevolence and compassion. Recognising the worth of others means recognising the worth of self in mutual recognition of one's humanity.

Fox Brindley and Ivanhoe offer a tempting vision of autonomy replaced by a dignity of agency and by the cultivation of a dual self, one social and one individual, which exist in balance. I will return to these ideas and discuss them further in later chapters.

---

[25] Hui E (2004) suggests the concept of Asian personhood differs from the Western person as 'thing in itself', the Asian version of a person being 'thing in relation to others'.

3.2.2.4   Attitudes to S&T in China

Miller (1996) argues that science in China has a subversive quality, but in fact the narrative is a conventional one of technological impetus and of China's global competitiveness.[26] How strongly does the Chinese public adhere to this view that S&T are 'always good'? The first survey of Chinese attitudes to S&T was undertaken in Beijing in 1989 and showed that the majority surveyed respected S&T as major contributors to economic success and the solving of China's issues, be they health, pollution, and so on (Zhang 1991). This is similar to a survey of attitudes towards nanotechnology conducted in Japan in 2004, where 85.6 % of respondents hoped that nanotechnology would benefit healthcare, and over 80 % said that it would solve environmental problems. Additionally, 98 % believed it would be of positive benefit, compared to 70 % in the UK and 68 % in the USA (Fujita 2006).

A survey of Chinese public scientific literacy, conducted in September 1990, demonstrated a need for greater public communication by scientists, a call supported by then president Hu Jintao in 2008 (Zhang 1993).[27] It is therefore unsurprising that a survey on GM food concluded that public non-awareness of the subject was widespread (Ma and Yandong 2009). Public perceptions on nanotechnology are low; according to Ma and Liao, a recent survey showed that when asked about nanotechnology, '60.7 % responded "'don't know'"', or have a wrong understanding'. Ma and Liao (2012) note that the discussion of nanotechnology in China was somewhat propaganda driven until 2003, after which 'the discussion became more objective and rational'. As Ma (2012) also notes,

> there are also some similarities between the Chinese and European publics: science has a very important place in the Chinese people's heart, and the way Chinese people access information is also becoming increasingly similar to that of the Europeans. In terms of perceptions of science and technology, the Chinese public holds a more optimistic view about science than the Europeans. Although the Chinese public is becoming more cautious and objective about science in recent years, such an attitude has not yet translated into the kind of worries and fears for science as seen in Europe. Science is still largely viewed as an important tool for transforming the world in the eyes of the Chinese public, and little attention is given to the possible challenges science may bring to our beliefs and values.

The following table sums up this differing approach in terms of a broader EU interest in the social impact of new technologies versus the more Chinese belief in innovation's economic power as primary (Table 3.2).

---

[26] Miller argues for a 1980s liberalization of science although notes that it is difficult to tell how wide ranging or indeed coherent this was; his own bias being for science as having an anti-authoritarian ethic is clear. He notes interestingly however the views of liberal scientists such as Fang Lizi and Xu Liangying in their 'open letters' of 1989.

[27] Also Liu et al. (2011).

Table 3.2 Differing bioethics, differing social contexts (China and the EU)

	EU	China
Social values	Dignity, freedom, democracy, equality, the rule of law, and respect for human rights	Family, harmony, community; *ren* or *jen* ethics – i.e. care for others, benevolence, humanness
Major cultural value, *difference in*	Individual human rights/autonomy a priority	Community prioritised over individual autonomy
Attitude towards S&T and scientists	Public differentiation between technologies, i.e. differing risks for nuclear, nano, GM	Innovation is always good
	Risk vs. benefit	*Tao* governs technology
	Post-Fukushima emphasis on confidence/or lack thereof in government, according to how it handles S&T crises	Harmony between *tian* (nature) and *ren* (human)
	Survey in 2010: 6 out of 10 Europeans felt that S&T can sometimes damage people's moral sense; 1 in 2 that applications of science and technology can threaten human rights	Trust in government (although this may be starting to decrease due to food scandals)
		Concern with social responsibility of scientists

3.2.3   Global Bioethics/Global Nanoethics?

> The importance of cultural diversity and pluralism should be given due regard. However, such considerations are not to be invoked to infringe upon human dignity, human rights and fundamental freedoms, nor upon the principles set out in this Declaration, nor to limit their scope. (UNESCO 2005)

Seven broad areas of global bioethics debate can be identified:

- Normative bases appropriate for the emerging field of global bioethics
- Global research ethics and what is owed to research subjects
- Collaborations
- Practices in education
- Euthanasia
- Global bioethics and religion
- Processes being developed among international organisations for identifying universal norms and values[28]

The above suggests common practices, rather than a set of common values. Dwyer (2003), for example, perpetuates that view when he argues not for a new set of concepts to address issues in global health but for detailed case studies instead.[29]

---

[28] On research subjects, see Macklin (2008); on collaborations, see Meslin (2008); and on processes, see Harris (2008).

[29] Dwyer notes the difference between Singer and Rawls in that the latter looks at laws and institutions and Singer at fundamental human interests – Rawls distinguishes the duty to assist citizens

Transformative steps for achieving global bioethics have been suggested based on processes of international knowledge sharing (Singer et al. 2006). The International Science and Bioethics Collaborations (ISBC) project is an example that brings together social anthropologists from Cambridge, Durham, and Sussex Universities as well as research partners from nine Asian countries. The project aims to address current social, economic, and cultural issues in international collaboration and knowledge exchange around bioscientific research, biomedicine, and bioethics. As Schummer (2006) notes, avoiding any cultural imperialism in bioethics is usefully achieved through international discussions:

> International discussion of ethical issues of nanotechnology is an excellent and important exercise, not only because views on nanotechnology are so diverse, but also because nanotechnology is frequently attached to a particularly strong and naïve attitude of 'improving the world'. International discussions can help us understand that our notions of both 'improvement' and 'the world' are very complex, culturally diverse and under continuous revisions.

And not merely international, but international *and* local, as per Maclurcan's (2009) caution against any nanotechnology approach that ignores the 'potential for local, village development of "appropriate' nanotechnologies", and that may even 'perpetuate' deficit thinking within international health, technology and development policy'.

There is no doubt that practical collaborations in bioethics distribute knowledge, arguably increase standardisation, and may offer more unified approaches to ethical issues, not merely in terms of global inclusion but also of multidisciplinarity – which may aid in those 'creative searches' for new methodologies that nanotechnology might require. However, these collaborations do not address the fundamental issues. Would work on global values underpin a global bioethics? How could we reconcile the dignity/human rights discourse in the West with the principle of 'harmony' espoused in the East? Are we arguing for a rather 'weak' philosophy that might combine Confucian ethics with Western rationalism and individualism? (Qui 2002)

3.3 Which *Values* for a Global Bioethics? Or Should We Consider Processes?

A 1993 International Bioethics Survey across ten countries suggested both a positive attitude to S&T and that the respondents are bioethically mature (Macer 1994).[30] Is this sufficient as a basis for global cooperation on a new technology such as nano?

---

of other countries from the principle of distributive justice that applies to the society of which one is a citizen.

[30] This survey however noted 'significant differences in public opinion concerning biotechnology in India, Thailand (and China) which poses a dilemma for policy makers'. Macer identifies greater

An article in *The Economist* on October 2, 2010, suggested that the 'philosophical question of whether universal values exist has turned into a political fight, dividing scholars, the media, and even, some analysts believe, China's leaders'. In China, the debate on universal values (*pushi jiazhhi*) allegedly began in 2008 after the Sichuan earthquake; in December of that year, 'Charter 08' was signed in support of universal values by several dissidents. However no clear outcome has emerged from that discussion.

One widely adopted approach in bioethics has been that of principlism, i.e. of Beauchamp's and Childress's argument in favour of the four principles of autonomy, beneficence, non-maleficence, and justice. Tangwa (2009) argues that these four principles are cross-culturally valid (using the particular example of Africa), in that 'although the emphasis given to each and the way they are applied or operationalized may differ... they form the basis of the similarities underlying the remarkable diversity of the sub-cultures'. Tai and Chung (2001) have commented that the principles of beneficence and non-maleficence have been a part of Eastern values since the time of Confucius.

Given differing Eastern and Western approaches to the idea of precaution, however, these principles may not be quite as unifying as we might like. Let us see if there might not be a more creative approach. One recent suggestion has been that of the idea of 'global solidarity' itself – the approach becomes the principle, as it were. Global solidarity promotes global equity as well as the five values of 'respect for human life; the intertwined relationship between human rights, responsibilities and needs; ensuring freedoms; democratic principles of accountability, representation, cooperation, and good governance; and environmental sustainability' (Baylis 2008). This seems to be a tall order.

Sakamoto (1999) gets to the heart of the issue when he discusses the idea of harmony, not solidarity. What he argues is that we need a Western agreement that is communitarian, or holistic – however one wishes to label it, a communal approach is required. This implies a Western recognition of *agreement* as more important than *individualism* per se:

> This work is the most crucial part of Global Bioethics, which is expected to harmonize and to bridge over all kinds of global ethoses, East and West, North and South. In this sense the new Global Bioethics should be 'holistic' in contrast to the 'individualistic' European model. Even today, Taoism, Confucianism, and Buddhism have a dominant influence on the ethos of the Asian world at its foundations... new bioethical issues... necessarily require some sort of communitarian way of thinking from the global point of view.

In other words, rather than arguing about values that we might agree on, we really need to be arguing about how we agree or disagree. Sakamoto does not simply offer this idea of harmony as the guiding principle of global bioethics but rather suggests the means for achieving it.

Sakamoto (2002a, b) argues for four key aspects in any possible global bioethics: (1) a new humanism, the (2) minimisation of human rights, (3) a holistic harmony,

---

enthusiasm for 'enhancement' therapies in these three countries than in the others surveyed. See also Patra et al. (2010).

and (4) the policy of global bioethics as a 'social tuning' technology. The second and third points obviously suggest some compromise between Western and Eastern narratives. The first proposes, albeit vaguely, 'a new philosophy concerning the relation between nature and the human being', but the fourth is arguably the most interesting of these ideas. Sakamoto (2002a, b) introduces the idea of policy to harmonise differing aspects of global bioethics based on 'bargain consensus', or 'dialogue bargain policy':

> I suggest that the global bioethics of the new century should not refer to any kind of 'universal principle', 'justice', 'categorical imperative', or the like for its policy. The only policy possible here will be continuing dialogue without any reference to any rigid principle, or, on the contrary, with reference to all antagonistic principles as impartial bargain alternatives, ultimately soothing the opposition and antagonism among the principles to reach a 'consensus of any kind'... We can imagine here that, through the process of reaching a bargain consensus, some sort of common feeling or compassion between both parties would be effectively born. I tentatively assume the existence of such a common feeling in all people as the 'Feeling of Ache and Pity'... I assume that we could best reconstruct a new, post-modern humanism based on this feeling of 'ache and pity'; for the third millennium.[31]

Sakamoto's dialogue draws on compassion and feelings of humanity, which may be similar to Tao's 'ethics of just caring', that allows, for context, relationship and particularity to counteract the impartiality of the reductionist approach to bioethics that she sees as currently prevalent.[32] Sakamoto's idea, in short, is that the *practice* of global dialogue will lead to the *principle*, or 'common feeling', that might advance a global bioethics.

Similarly, Hongladarom (2004) argues that 'decisions as how people from different backgrounds are to co-exist with one another peacefully' should be made on an 'overlapping consensus' which is 'shorn of the metaphysical basis on what constitutes the good life of the respective groups that enter into the deliberation'. In line with this argument, this study will look at the *practices* of global ethics – particularly dialogue/participatory technology assessment – rather than venturing into the realm of universal values. We will thus begin with practice on a global policy level.

In conclusion, the bioethics context is different in the East and the West. As personhood can only be acquired through social practice, according to Confucian teachings, human value evolves from an individual's social relations. Confucianism describes each human being as a bearer of specific social roles: as father, as son, as wife, as ruler, as subject, and as friend. Each role is assigned a different status and a different pattern of behaviour or duties. This is not a system of individual rights but of roles (von Senger 1993).

Western individualism takes a different approach, one more based on the right to an individual's autonomy. This outlines the basic conflict that is often perceived in broad terms between the West and the East. However, this chapter makes the following two claims: first, that 'dignity', seen as a primary value in Western discourse on

---

[31] See also Sakamoto (2004).

[32] However she also suggests that a 'just caring' ethics might require some form of moral framework and would need to be grounded in some conception of a good life.

the individual and his or her rights, is a more universal quality than is usually thought and, second (and as a related point), that agency is a preferred term to autonomy or dignity. Dignity should not even be seen as a form of autonomy, as that too has connotations of human rights, but of virtuous agency. This suggests that it might more effectively be seen as a process, rather than a values-based system.

References

Alora RAT, Lumitao JM (eds) (2001) Beyond a western bioethics: voices from the developing world. Georgetown University Press, Washington, DC

Andorno R (2007) First steps in the development of an international biolaw. In: Dierickx K, Nys H, Schotmans P (eds) New pathways in European bioethics. Intersentia, Antwerpen, pp 121–138

Angel S (2002) Human rights in Chinese thought: a cross-cultural inquiry. Cambridge University Press, Cambridge

Arekapudi S, Wynia M (2003) The unbearable whiteness of the mainstream: should we eliminate, or celebrate, bias in bioethics? Am J Bioeth 3:18–29

Barr MD (2002) Cultural politics and Asian values: the Tepid War. Routledge, London

Baylis F (2008) Global norms in bioethics: problems and prospects. In: Green RM, Donovan A, Jauss SA (eds) Global bioethics. Issues of conscience for the twenty-first century. Clarendon, Oxford, pp 324–341

Beauchamp TL, Childress JF (2001) Principles of biomedical ethics. Oxford University Press, New York

Bell DA (2006) Liberal democracy: political thinking for an East Asian context. Princeton University Press, Princeton

Bielefeldt H (1995) Muslim voices in the human rights debate. Human Rights Q 17(4):587–617

Callahan D (2002) Bioethics. In: Chadwick R, Schroeder D (eds) Applied ethics: critical concepts in philosophy, vol II. Routledge, London, pp 3–19

Changthavorn T (2003) Bioethics of IPRs: what does a Thai Buddhist think? Paper presented at the roundtable discussion on Bioethical Issues of IPRs, Selwyn College, University of Cambridge

Chattopadhyay S (2008) Bioethical concerns are global, bioethics is Western. Eubios J Asian Int Bioeth 18(July):106–109

China Daily (2011) Different cultures show same respect for human dignity. http://usa.chinadaily.com.cn/opinion/2011-09/22/content_13764204.htm. Accessed 22 Dec 2014

Chu Y, Diamond L, Nathan AJ, Shin DC (2009) How East Asians view democracy. Columbia University Press, Columbia

Council of Europe (1997) Convention for the protection of human rights and dignity of the human being with regard to the application of biology and medicine: convention on human rights and biomedicine. http://conventions.coe.int/Treaty/en/Treaties/Html/164.htm. Accessed 22 Dec 2014

De Bary WT (1998) Asian values and human rights. A Confucian communitarian perspective. Harvard University Press, Cambridge

Dwyer J (2003) Teaching global bioethics. Bioethics 17(5/6):432–446

Ebbesen M (2007) Nanoethics – not from scratch. Paper presented at the nano ethics workshop, University of Aarhus, Denmark, 22–23 September 2007

Emerson DK (1995) Singapore and the "Asian values' debate. J Democr 6(4):95–105

Eurobarometer (2010) Science and technology report. http://ec.europa.eu/public_opinion/archives/ebs/ebs_340_en.pdf. Accessed 22 Dec 2014

European Commission (2010) Report on the European Commission's public online consultation. Towards a strategic nanotechnology action plan (SNAP) 2010–2015. http://ec.europa.eu/research/consultations/snap/report_en.pdf. Accessed 22 Dec 2014

Fan R (1997) Self-determination vs. family determination: two incommensurable principles of autonomy. Bioethics 11(3/4):309–322

Fan R (1998) Truth-telling to the patient: cultural diversity and the East Asian perspective. In: Fujiki N, Macer D (eds) Bioethics in Asia. Eubios Ethics Institute, Tsukuba, pp 107–109

Fan R (2006) Bioethics: globalization, communitization, or localization? In: Engelhardt HT Jr (ed) Global bioethics: the collapse of consensus. M&M Scrivener Press, Salem, pp 271–299

Feldman D (1999) Human dignity as a legal value – Part 1. Public Law (Winter):682–702

Fox Brindley E (2010) Individualism in early China, human agency and the self in thought and politics. University of Hawai'i Press, Honolulu

Fujita Y (2006) Perception of nanotechnology among general public in Japan. Asia Pac Nanotechnol Wkly 4(6). http://www.nanoworld.jp/apnw/articles/library4/pdf/4-6.pdf. Accessed 22 Dec 2014

Gaskell G, Allum N, Stares S (2003) Europeans and biotechnology in 2002. http://ec.europa.eu/public_opinion/archives/ebs/ebs_177_en.pdf. Accessed 22 Dec 2014

Gbadegesin S (2007) The moral weight of culture in ethics. In: Pellegrino ED, Prograis LJ (eds) African American bioethics: culture, race, and identity. Georgetown University Press, Washington, DC, pp 24–45

Gert B, Culver CM, Clouser KD (1997) Bioethics: a return to fundamentals. Oxford University Press, Cary

Giles J (2004) Nanotechnology: what is there to fear from something so small? Nature 1:43

Grunwald A (2005) Nanotechnology – a new field of ethical enquiry. Sci Eng Ethics 11:187–201

Guo C (2011) Conceiving conception: the bioethics of assisted reproductive policy in China. http://www.wcfia.harvard.edu/node/6535. Accessed 22 Dec 2014

Guston D (ed) (2010) Encyclopedia of nanoscience and society, vol II. Sage, Thousand Oaks

Harbour FV (1999) Thinking about international ethics. Westview Press, Boulder

Harris J (2008) Global norms, informed consensus, and hypocrisy in bioethics. In: Green RM, Donovan A, Jauss SA (eds) Global bioethics. Issues of conscience for the twenty-first century. Clarendon, Oxford, pp 297–323

Hennig W (2006) Bioethics in China: although national guidelines are in place, their implementation remains difficult. EMBO Rep 7(9):850–854

Hongladarom S (2004) Asian bioethics revisited: what is it? And is there such a thing? Eubios J Asian Int Bioeth 14(6):194–197

Hu W, Chen K (2004) Chinese consumers be persuaded? The case of genetically modified vegetable oil. AgBioForum 7(3):124–132. http://www.agbioforum.org. Accessed 22 Dec 2014

Hui E (2004) Personhood and bioethics: Chinese perspective. In: Qui R (ed) Bioethics: Asian perspectives – a quest for moral diversity. Kluwer Academic Publishers, Dordrecht/Boston, pp 29–44

Ip M, Gilligan T, Koenig T, Raffin J (1998) Ethical decision-making in critical care in Hong Kong. Crit Care Med 26(3):447–451

Ivanhoe PJ (2000) Confucian moral self-cultivation. Hackett, Indianapolis

Jia L, Zhao Y, Liang XJ (2011) Fast evolving nanotechnology and relevant programs and entities in China. Nano Today 6(1):611

Joseph T, Morrison M (2006) Nanotechnology in agriculture and food. ftp://ftp.cordis.europa.eu/pub/nanotechnology/docs/nanotechnology_in_agriculture_and_food.pdf. Accessed 22 Dec 2014

Kachonpadungkitti C, Macer D (2004) Attitudes to bioethics and biotechnology in Thailand (1993–2000), and impacts on employment. Eubios J Asian Int Bioeth 14:118–134

Kaelin L (2009) Contextualizing bioethics: the UNESCO declaration on bioethics and human rights and observations about Filipino bioethics. Eubios J Asian Int Bioeth 19(March):42–47

Keiper A (2007) Nanoethics as a discipline? New Atlantis (Spring):55–67

Kemp P, Rendtorff JD (eds) (2000) Basic ethical principles in European bioethics and Biolaw. Autonomy, dignity, integrity and vulnerability, vol I. Centre for Ethics and Law/Institut Borja de Bioètica, Copenhagen/Barcelona

Keys S, Ma JK, Drake PM (2008) Genetically modified plants and human health. J R Soc Med 101(6):290–298. http://jrsm.rsmjournals.com/cgi/content/full/101/6/290. Accessed 22 Dec 2014

Kim SY (2012) Do Asian values exist? Empirical tests of the four dimensions of Asian values. J East Asian Stud 10(2):315–344

Klein K (2010) Illiberal biopolitics and embryonic life: the governance of human embryonic stem cell research in China. http://www.hks.harvard.edu/sdn/articles/files/Klein-Illiberal%20Biopolitics.pdf. Accessed 22 Dec 2014

Lee S (1992) Was there a concept of rights in Confucian virtue-based morality? J Chin Philos 19:241–261

Lenoir N, Mathieu B (1998) Les norms interrnationales de la bioethique. Presses Universitaires de France, Paris

Li E (2008) Bioethics in China. Bioethics 22(8):448–454

Lin AC (2006) Size matters: regulating nanotechnology. http://ssrn.com/abstract=934635. Accessed 22 Dec 2014

Liu L (1993) Translingual practice: the discourse of individualism between China and the West. Positions 1(1):1–34

Liu Y, Duan Y, Xiao G, Tang L (2011) S&T policy evolution: a comparison between the United States and China (1950–present). Paper presented at the 2011 Atlanta conference on science and innovation policy

Ma Y (2012) Public perceptions of science and technology in China. In: Ethics state of the art: state of debate in the three regions, GEST report, May 2012 (supplied by Dr Miltos Ladikas)

Ma Y, Liao M (2012) Nano-technology development in China and three related ethic discussions. Presented at the GEST roundtable, CASTED, 5 September 2012, (Supplied by Dr. Miltos Ladikas)

Ma Y, Yandong Z (2009) Analysis of Beijing resident's satisfaction to food safety. Social Science of Beijing 3 (GEST project copy)

Macer D (1994) International bioethics survey – world view. In: Macer D (ed) Bioethics for the people by the people. Eubios Ethics Institute, Christchurch, pp 125–138

Macklin R (2003) Dignity is a useless concept. Br Med J 327:1419–1420

Macklin R (2008) Global justice, human rights, and health. In: Green RM, Donovan A, Jauss SA (eds) Global bioethics. Issues of conscience for the twenty-first century. Clarendon, Oxford, pp 141–160

Maclurcan DC (2009) Southern roles in global nanotechnology innovation: perspectives from Thailand and Australia. NanoEthics 3:137–156

Madsen R (2007) Confucianism: ethical uniformity and diversity. In: Sullivan WH, Kymlicka W (eds) The globalization of ethics. Religious and secular perspectives. Cambridge University Press, New York, pp 17–133

Meslin EM (2008) Achieving global justice in health through global research ethics: supplementing Macklin's "Top-Down" approach with one from the "Ground Up". In: Green RM, Donovan A, Jauss SA (eds) Global bioethics. Issues of conscience for the twenty-first century. Clarendon, Oxford, pp 163–178

Miller HL (1996) Science and dissent in post-Mao China. The politics of knowledge. University of Washington Press, Seattle

Miller G (2006) Nanomaterials, sunscreens and cosmetics: small ingredients, big risks. Friends of the earth Australia and USA report. http://libcloud.s3.amazonaws.com/93/ce/0/633/Nanomaterials_sunscreens_and_cosmetics.pdf. Accessed 22 Dec 2014

Minato K (2007) Cross-national social survey in East Asia. http://jgss.daishodai.ac.jp/research/monographs/jgssm7/jgssm7_14.pdf. Accessed 22 Dec 2014

Mitchell P (2003) UK government caught in GM dilemma. Nat Biotechnol 9(21):957

Moor J, Weckert J (2004) Nanoethics. In: Baird D, Nordmann A, Schummer J (eds) Discovering the nanoscale. IOS Press, Amsterdam, pp 301–310

Myser C (2003) Differences from somewhere: the normativity of whiteness in bioethics in the United States. Am J Bioeth 3(2):1–11

Nanowerk (2011) European food safety authority publishes nanotechnology guidance for food and feed assessment. http://www.nanowerk.com/news/newsid=21308.php. Accessed 22 Dec 2014

Nathan AJ, Chu Y, Myers JJ (2008) How East Asians view democracy. Carnegie Council podcast, November 2008. http://www.carnegiecouncil.org/resources/transcripts/0085.html. Accessed 18 Nov 2014

Park CM, Shin DC (2004) Do Asian values deter popular support for democracy? The case of South Korea. Asian Barometer working paper series: no 26. http://www.asianbarometer.org/newenglish/publications/workingpapers/no.26.pdf. Accessed 22 Dec 2014

Patra D, Haribabu E, McComas KA (2010) Perceptions of nano ethics among practitioners in a developing country: a case of India. NanoEthics 4(1):67–75

Petroni AM (2006) Perspectives for freedom of choice in bioethics and health care in Europe. In: Engelhardt HT Jr (ed) Global bioethics: the collapse of consensus. M&M Scrivener Press, Salem, pp 238–270

Pinker S (2008) The stupidity of dignity. New Repub 238(9):28–31

Pollard I (2007) High tech neuroscience, neuroethics, and the precautionary principle. In: UNESCO Bangkok Asia-Pacific perspectives on ethics of science and technology. UNESCO, Bangkok

Polo M (2011) Food and nano-food within the Chinese regulatory system: no need to have over-regulation. Eur J Law Technol 2(3):1–16

Pye LW (1996) The state and the individual: an overview interpretation. In: Hook B (ed) The individual and the state in China. Clarendon, Oxford, pp 16–42

Qiu R (1992) Medial ethics and Chinese culture. In: Pellegrino E, Mazzarella P, Corsi P (eds) Transcultural dimensions in medical ethics. University Publishing Group, Frederick, pp 159–180

Qiu R (2004) Bioethics and Asian culture. In: Qui R (ed) Bioethics: Asian perspectives – a quest for moral diversity. Kluwer Academic Publishers, Dordrecht/Boston, pp 34–52

Qiu R (2008) Bioethics in China (1990–2008). Asian Bioeth Rev 1(December):44–57

Qiu R (2011) Reflections on bioethics in China: the interactions between bioethics and society. In: Myer C (ed) Bioethics around the globe. Oxford University Press, Oxford, pp 164–190

Qui R (2002) The tension between biomedical technology and Confucian values. In: Tao JLP (ed) Cross-cultural perspectives: on the (im)possibility of global bioethics. Kluwer, Dordrecht, pp 71–88

Reich WT (1994) The word 'bioethics': its birth and the legacies of those who shaped it. Kennedy Inst Ethics J 4(4):319–335

Reich WT (ed) (1995) Encyclopedia of bioethics. Macmillan, New York

Rendtorff J (2003) Basic ethical principles in European bioethics and biolaw: autonomy, dignity, integrity and vulnerability – towards a foundation of bioethics and biolaw. Med Health Care Philos 5(3):235–244

Ryan MA (2004) Beyond a western bioethics? Theol Stud 65:158–177

Sakamoto H (1999) Towards a new global bioethics. Bioethics 13:191–197

Sakamoto H (2002a) Is just caring possible? Challenge to bioethics in the new century. In: Tao JLP (ed) Cross-cultural perspectives: on the (im)possibility of global bioethics. Kluwer, Dordrecht, pp 41–58

Sakamoto H (2002b) A new possibility for global bioethics as an intercultural social turning technology. In: Tao JLP (ed) Cross-cultural perspectives: on the (im)possibility of global bioethics. Kluwer, Dordrecht, pp 359–368

Sakamoto H (2004) Globalisation of bioethics – from the Asian perspective. In: Macer D (ed) Challenges for bioethics in Asia. The proceedings of the fifth Asian bioethics conference. Eubios Ethics Institute, Christchurch, pp 487–494

Sass HM, Xiaomei Z (2011) Global bioethics: Eastern or Western principles? Asian Bioeth Rev 3(1):1–2. http://muse.jhu.edu/journals/asian_bioethics_review/toc/asb.3.1.html. Accessed 22 Dec 2014

Schroeder D (2010) Dignity: one, two, three, four, five, still counting. Camb Q Healthc Ethics 19(1):118–125

Schroeder D (2012) Human rights and human dignity: an appeal to separate the conjoined twins. Ethic Theory Moral Pract 15:323–335

Schummer J (2006) Cultural diversity in nanotechnology ethics. Interdiscip Sci Rev 31(3):217–230

Schwartz BI (1985) The world of thought in ancient China. Harvard University Press, Cambridge, MA

Scott N (2009) Research ethics: European and Asian perspectives, global challenges. In: Ladikas M (ed) Embedding society in science & technology policy. European and Chinese perspectives. European Commission, Brussels

Shuang Y (2011) Anti-GM food protestors claim crops are unsafe. Global Times. http://www.globaltimes.cn/NEWS/tabid/99/ID/675011/Anti-GM-food-protesters-claim-crops-are-unsafe.asp. Accessed 22 Dec 2014

Singer PA, Bhatt A, Frew SE, Greenwood H, Mackie J, Panjwani D, Persad DL, Salamanca-Buentello F, Séguin B, Taylor AD, Thorsteinsdóttir H, Daar AS (2006) Harnessing genomics and biotechnology to improve global health equity. In: Green RM, Donovan A, Jauss SA (eds) Global bioethics. Issues of conscience for the twenty-first century. Clarendon, Oxford, pp 179–204

Solomon D (2006) Domestic disarray and imperial ambition: contemporary applied ethics and the prospects for global bioethics. In: Engelhardt HT (ed) Global bioethics: the collapse of consensus. M&M Scrivener, Sudbury, pp 335–361

Steiner H, Alston P (2000) International human rights in context. Oxford University Press, Oxford

Stonington S, Ratanakul P (2006) Is there a global bioethics? End-of-life in Thailand and the case for local difference. PLoS Med 3(10), e439

Sullivan WM (2007) Ethical universalism and particularism: a comparison of outlooks. In: Sullivan WM (ed) The globalization of ethics. Religious and secular perspectives. Cambridge University Press, Cambridge, pp 191–212

Tai MC, Chung SL (2001) Developing a culturally relevant bioethics for Asian people. J Med Ethics 27:51–54

Tangwa GB (2009) Ethical principles in health research and review process. Acta Trop 112S:S2–S7

The Economist (2010) The debate over universal values, October 2, 2010, pp 65–66

The Hastings Center (2015) About us. http://www.thehastingscenter.org/About/Default.aspx. Accessed 22 Dec 2014

The National Foreign Trade Council (2003) EU regulation, standardization, and the precautionary principle. http://www.wto.org/english/forums_e/ngo_e/posp47_nftc_eu_reg_final_e.pdf. Accessed 22 Dec 2014

The New Atlantis (2003) Chinese bioethics? "Voluntary" eugenics and the prospects for reform. New Atlantis 1:138–140

Thompson MR (2001) Whatever happened to Asian values? J Democr 12:154–165

Traer R (1991) Faith in human rights: support in religious traditions for a global struggle. Georgetown University Press, Washington, DC

UNESCO (2005) Declaration on bioethics and human rights, Article 12. www.undp.org/hdr2001. Accessed 22 Dec 2014

University of Cambridge (2014) Partners linked across collaborations in ethics and biosciences. http://www.placebo.socanth.cam.ac.uk/isbcsubproj2b.php. Accessed 22 Dec 2014

van Calster G (2008) Risk regulation, EU law and emerging technologies: smother or smooth? NanoEthics 2:61–71

von Senger H (1993) Chinese culture and human rights. In: Schmale W (ed) Human rights and cultural diversity – Europe, Arabic-Islamic world, Africa and China. Keip Publishing, London, pp 295–309

Wang Y (1997) AIDS, policy and bioethics: ethical challenges facing China in HIV prevention. Bioethics 11(3/4):323–326

Wang Y (2003) Chinese ethical views on embryo stem cell research. Asian bioethics in the 21st century. Eubios Ethics Institute. http://eubios.info/ABC4/abc4049.htm. Accessed 22 Dec 2014

Wang Y (2011) Group protection in human population genetic research in developing countries: the People's Republic of China as an example. University of Glasgow. http://theses.gla.ac.uk/3005/. Accessed 22 Dec 2014

Weckert J, Moor J (2006) The precautionary principle in nanotechnology. Int J Appl Philos 20(2):191–204

Wertz DC, Fletcher JC, Berg K (2003) Review of ethical issues in medical genetics. World Health Organization. http://apps.who.int/iris/bitstream/10665/68512/1/WHO_HGN_ETH_00.4.pdf?ua=1. Accessed 22 Dec 2014

Widdows H (2011) Western and Eastern principles and globalised bioethics. Asian Bioeth Rev 3(1):14–22

Xinhua (2010) Debate on GM food continues. China Daily. http://www.chinadaily.com.cn/china/2010-03/04/content_9534076.htm. Accessed 22 Dec 2014

Xinhua (2012) GM food testing worries parents. China Daily. http://www.chinadaily.com.cn/china/2012-09/12/content_15753932.htm. Accessed 22 Dec 2014

Xuequan M (ed) (2010) GM food: hope or fear for the Chinese? http://news.xinhuanet.com/english2010/china/2010-10/16/c_13559695.htm. Accessed 22 Dec 2014

Yang X (2004) An embryonic nation. Nature 428:210–212

Yu K (2003) The alleged Asian values and their implications for bioethics. Eubios. http://www.eubios.info/ABC4/abc4232.htm. Accessed 22 Dec 2014

Zhang Z (1991) People and science: public attitudes in China toward science and technology. Sci Public Policy 18(5):311–317

Zhang Z (1993) A survey of public scientific literacy in China. Public Underst Sci 2(1):21–38

Zhang Q (2000) The idea of human dignity in classical Chinese philosophy: a reconstruction of Confucianism. J Chin Philos 27(3):299–330. http://article.chinalawinfo.com/Article_Detail.asp?ArticleId=32900. Accessed 22 Dec 2014

Žižek S (2003) Bring me my Philips Mental Jacket. London Review of Books. http://www.lrb.co.uk/v25/n10/slavoj-zizek/bring-me-my-philips-mental-jacket. Accessed 22 Dec 2014

Chapter 4
Nanoregulation

The European Commission is losing the leadership on nano. We cannot support innovation whilst disregarding the protection of human health and environments. Without the citizens' support, the nano future is condemned (Nanowerk 2012).

...it is still necessary to set up a viable system based on the precautionary principle... since research in the field of social and ethical evaluation of nanotechnology developments in China is not as advanced as in the USA and Europe, cooperation is important to avoid making similar mistakes and to promote the smooth development of nanotechnology (Decker and Li 2009).

Abstract This chapter compares regional policies and regulations in China and the EU such as REACH and EU-CoC, before examining attempts at global regulation through bodies such as the OECD Working Party on Nanotechnology (WPN) and the OECD Working Party on Manufactured Nanomaterials (WPMN).

Keywords Regulation • China nanopolicy • EU-CoC • REACH

In 2007, Kjolberg and Wickson offered the view in the journal *NanoEthics* that current nanoregulation is insufficient and that more research on toxicology was needed. They identified regulation as the first of the four major concerns raised by commentators up to that date. This raises three questions, which this chapter will address:

1. What *are* the risks of nanotechnology in terms of the impact on human health and the environment, as seen by commentators in China and in the EU?
2. What regulation is currently in place in these two regions for nanotechnology? (Is it a new regulation or adapted?)
3. What *global* regulation is currently in place for nanotechnology?

S. Dalton-Brown, *Nanotechnology and Ethical Governance in the European Union and China*, DOI 10.1007/978-3-319-18233-9_4

4.1   What Are the Risks, and What Regulatory Approaches Might We Take?

Given, as discussed in Chap. 3, the view of nanotechnology as a potentially runaway technology that can be used by 'dark forces', what exactly are the risks (as perceived by researchers)?

The effect of nanotoxins has yet to be fully determined, and lab studies have shown that there might be some respiratory system damage through the inhalation of nanoparticles, leading to the fear that nanotech might potentially be 'the next asbestos' (Anderson et al. 2009). Andrew Maynard (2007), chief scientist for the Project for Emerging Nanotechnologies, suggested in 2006 a need to 'energize' the global research community to tackle potential nanotechnology dangers. In his testimony before a US federal government committee, he asked the following questions:

> What effect do airborne nanoparticles have on the lungs? Do nanoparticles penetrate the skin? What happens to nanoparticles in water? How do they behave in the gastrointestinal tract? What happens to nanoparticles when they are poured down a drain and enter the waste stream?

4.1.1   Risks

Nanomedicine, due to the action of particles about which not enough is yet known, may cause cardiovascular disease, increasing the risk of heart attacks and strokes (Fernholm et al. 2008),[1] with some critics also arguing that 'there is no question that the invasion of cells by nanoparticles could be carcinogenic' (Motavalli 2010) and that some nanoparticles already in commercial use are toxic to cells. These can allegedly damage DNA, negatively affect proteins, and cause cell death (Gerloff et al. 2009; Hussain et al. 2010). A study at the University of Rochester found that when rats breathed in nanoparticles, the particles settled in the brain and lungs, which led to significant increases in biomarkers for inflammation and stress response (Buzea et al. 2007). Mice studies have also found that nanoscale titanium dioxide can cause genetic instability and can pass from pregnant mice to their offspring, damaging their genital and cranial nerve systems (Tsuchiya et al. 1996). (Titanium dioxide is used in allegedly close to 400 sunscreen products.) A study in China indicated that nanoparticles induce skin aging in hairless mice, while a 2-year study at UCLA's School of Public Health found laboratory mice that had consumed nanoscale titanium dioxide showed DNA and chromosome damage (Elder 2006).[2] Several types of engineered nanomaterials including titanium dioxide and carbon

---

[1] See also Satariano (2010) and Evans (2007).
[2] See also Wu et al. (2009).

nanotubes are believed to produce pulmonary inflammation and fibrosis in animals (Shvedova et al. 2009) and[3] even brain damage (Oberdorster 2004).[4]

Andre Nel (Chief Scientist, Nanomedicine, Los Angeles' NanoSystems Institute at the University of California) has discussed the need to assess whether the 'nano-carriers' that transport drugs have:

> hazardous effects that are different and independent from the drugs being delivered… so far the only studies on the effects of nanotechnology in animals have focused on industrial nanomaterials rather than those used in nanomedicine (and) the same screening methods will be used to look at the safety of nanodrugs. (Makoni 2010)

Som et al. (2010) also warn that health risks associated with indirect exposure of humans to nanoparticles in the environment cannot be ignored. They give the cautionary example of children facing harmful lead exposure through the intake of soil and dust contaminated by lead-based paints falling off walls and facades.

There is no clear resolution to the issue of nanotechnology's environmental effects. One problem is that of the length of time that nanoengineered materials may last in landfill sites. Thus, a 2003 Greenpeace survey suggested that the environmental impact of a new class of non-biodegradable pollutants needed consideration (Arnall 2003).

A 2010 Friends of the Earth report suggested that many promised benefits from nanotechnology in terms of improving energy capture, manufacture and storage are counteracted by the high-energy demands and environmental impacts of manufacturing nanomaterials. Nanoenhanced solar panels are not only energy inefficient to manufacture, they appear to have a considerably reduced lifespan in comparison with conventional panels. Therefore, even if the manufactured products may be more energy efficient, the production cost and the lack of durability will diminish that saving. And besides, as the report gloomily states, environmental products are a tiny section of a market that focuses more on commercial applications:

> Most nanoproducts are not designed for the energy sector and will come at a net energy cost. Super strong nano golf clubs, wrinkle disguising nanocosmetics, and colour-enhanced television screens take a large quantity of energy to produce, while offering no environmental savings. Such nanoproducts greatly outnumber applications in which nano could deliver net energy savings. (Illuminato and Miller 2010)[5]

A UNESCO 2006 report referred to the risk of scientists being 'no longer capable of autonomously directing scientific research due to the growth of external pressures'. One assumes that these pressures include the hunger for biotech advances and the tremendous commercial potential they allegedly represent, as the report goes on to discuss the allegedly overzealous granting of patents (UNESCO 2006a, b).

Yet corporate interest in nanotechnology is not the issue, corporate corner-cutting in the rush to market, and of global access to products, is. Industry agents may, in

---

[3] See also Avolainen et al. (2010).

[4] See also Holmes (2004).

[5] See also Berger (2012).

such a rush to market, 'realise that it is in their interest to present nanotechnology as "business as usual" or "evolutionary" in order to render nanotechnology familiar and therefore harmless' (Shelley-Egan 2010). The opposing view can, of course, be put that without corporate funding for the major potential applications of nanotechnology, many beneficial products would not be manufactured (Hassan and Sheehan 2003).

McCray (2009), looking at the seminal work of Roco in promoting nanoresearch as a US national priority, notes a particular ethical concern about how the defence industry has played a significant part in funding these advances.

Nanotechnology has been condemned for its potential to advance Western consumerism, and a little yet has been aimed at products that might benefit the poor (McKibben 2003).[6] Whereas nanotechnology is an important driver for economic success, the West has some concerns that nanoproducts may lead to a displacement of jobs and major changes in trade balances between countries; manufacturing in countries with weaker controls (who subsequently export worldwide) may increase risk, and benefits may be unevenly distributed (leading to a so-called global 'nanodivide' between North and South).

In conclusion, the risk of toxicity to human health and the environment remains uncertain in the long term, while economic so-called risk can often be seen as a positive. It can encourage market responsibility on the part of corporations sensitive to the threat of consumer activism. In the EU health and environment are the two (interrelated) major areas of risk, with many countries, particularly in Europe, remembering the introduction of drugs such as thalidomide in the 1960s, as well as the GM food debate. The tendency towards adopting the precautionary principle means that the EU is reasonably risk averse.

The situation is different in China, where the precautionary principle is not pervasive, yet China is catching up in terms of nanosafety concerns. The issue for China has intensified after a nanoparticle exposure accident in a poorly ventilated Chinese paint factory in August 2007. Seven workers contracted lung disease, and 2 of those died (Song et al. 2009). After this incident, 'Chinese policymakers shifted focus to the risk management aspects of nanotechnology', an example of which is the large-scale program begun in 2011 on monitoring factory worker exposure (Tang et al. 2011). The 2007 incident raised several questions about whether the link between exposure and pulmonary illness can be proven. There are now more than 30 research organisations in China that have initiated research activities studying the toxicological and environmental effects of nanomaterials and nanoparticles, as well as a further 120 research organisations that are undertaking general research into nanoscience and nanotechnology (Zhao et al. 2008).[7]

---

[6] See also Mnyusiwalla et al. (2003).

[7] See also Cheng et al. (2009).

4.1.2   Four Types of Regulatory Approaches

Before looking at the differing regulatory approaches in each region, we might first briefly note that the word 'regulation' can mean different things. This is significant in terms of whether regulation is 'open' (i.e. involving a mixture of governmental, NGO and civil engagement) or closed, meaning purely top-down in its approach. There are four types of regulatory initiatives:

1. Registers such as the UK Voluntary Reporting Scheme and the Swiss Nano-inventory
2. Risk management systems, ranging from governmental such as the NanoKommission, for example, or business generated, such as the Cenarios (Certifiable Nanospecific Risk Management and Monitoring System) system introduced in Germany in 2008
3. Codes of conduct generated by governmental agencies, NGOs or business initiatives (business/NGO partnerships), such as the IG-DHS Code of Conduct for Nanotechnology or the BASF Code of Conduct Nanotechnology
4. Actual regulatory policy, which is legally enforceable. Currently no specific nanoregulation exists – nanorisk is covered by existing legislation. However, the first international law designed specifically for nanotechnology is coming into effect as this study is being completed (July 2013). This is regulation (EC) no 1223/2009 of the European Parliament and of the council of 30 November 2009 on cosmetic products, which will require label notification of 'the presence of substances in the form of nanomaterials', according to Article 13 (Nanowerk 2013).[8]

The use of the word '*voluntary*' gives the reader an idea of why the first type of nanoregulation is considered ineffective. A 2004 study of public perceptions of nanotechnology in the USA showed that 60 % of respondents did not trust business leaders to minimise nanotechnology risks (Cobb and Macoubrie 2004). Yet there is the increased power of the consumer. The GM food scandal and the economic results of the GM boycott (discussed in Chap. 2) indicate that corporations need to be careful in how their markets perceive risk. This is an issue for both China and the EU (Applebaum and Parker 2007).

The 2006 'Voluntary Reporting Scheme' introduced in the UK in terms of which companies would report to the UK Department for Environment, Food and Rural Affairs has been negatively assessed for a lack of compliance incentives. In 2008, the US Environmental Protection Agency (EPA) implemented a program for the nanotech industry to report their use of nanomaterials voluntarily – a program that also appears to have failed (Allhoff et al. 2010).[9] However, Allhoff et al. (2010), in their consideration of what effective regulation might look like, suggest that a compromise solution involving nanoindustry self-regulation plus legislation may be the

---

[8] 'For the actual regulation, see Office Journal of the European Union (2009).
[9] See also Trager (2008).

best solution. The evolving context surrounding regulation can include hybridity of civil and governmental regulation, with national or state codes plus non-conventional, voluntary, private-sector-initiated regulatory arrangements (Bowman and Hodge 2008).

Risk management systems have also been criticised for their voluntary nature. Cenarios, however, described by the company as the 'first and only nanosafety standard with certificate worldwide', has an external body that certifies nanosafety.

Any moves towards transparency of this kind create a new global culture of awareness, as well as pressuring governments towards greater action on nanorisk.

One might mention in this context the role of environmental NGOs. The response to the European Commission's second regulatory review of nanomaterials by a group of NGOs, the European Environmental Bureau and various consumer organisations and trade unions on October 23, 2012, has been strongly critical of the lack of governmental action on nano-specific regulation, stating that:

> The European Commission is losing the leadership on nano. We cannot support innovation whilst disregarding the protection of human health and environments. Without the citizens' support, the nano future is condemned. (Nanowerk 2012)

The BASF *Code of Conduct* (deriving from the German chemical corporation BASF) is tied to the company's image and marketing the company as responsible.[10] However, the IG-DHS (an acronym referring to a group of Switzerland's biggest retailers) Code, introduced in 2008, is described as a 'strong measure', in that it demands precise information from suppliers as to nanomaterials in their products (Fiedeler et al. 2010). The IG-DHS suggests, as mentioned under point 2 above, that due to a lack of appropriate regulation at a governmental level, private companies' codes of conduct have become more important. Thus, the UK has a NanoCode, the Nanocare Initiative and the NanoRisk Framework (Ferrari 2010).

In terms of both regulation and policy, the issue is whether existing *policies* cover nanorisks sufficiently or whether, as in the USA, where some nanomaterials can be regulated within the existing Toxic Substances Control Act, this may be neither ideal nor sufficiently far-reaching (Preston et al. 2010). The seminal 2004 Royal Society Report in the UK accepted current regulatory frameworks in that country as sufficient but recommended a review and caution given the lack of evidence as yet on the effect of, for example, inhaled nanoparticles, as well as the 'avoidance as far as possible' of emitting nanoparticles. A 2004 EU legislative review – launched for assessment of nanorisk and need for regulatory intervention – offered 12 recommendations, such as a new nomenclature for nanochemicals to highlight that these are in fact *new* (thus their effects cannot necessarily be considered similar to their non-nano versions), better risk assessment mechanisms, greater data and so forth. The three major conclusions, across a range of papers on risk in this review, were:

---

[10] See BASF (2013) for the 'Dialogforum Nano of BASF 2011|2012 Transparency in communication on nanomaterials from the manufacturer to the consumer' report.

- That there is as yet not enough data on the risks of nanoproducts.
- Given the above, there is a need to improve risk assessment mechanisms and criteria, to be cautious about nanoparticle elimination into the environment.
- There is a need for dialogue (the EU report lists this as the 8th of its 12 recommendations) (European Commission 2004).

Does nanotechnology require new, specific regulatory approaches? There is no particular agreement on what a new system of regulation should look like, or how it might be achieved, with new risk assessment tools, greater public participation and new global codes of conduct being just some of the pathways suggested (Hunt 2008). As Guerra (2008) states, the task is not easy:

> Major issues arise when one tries to answer the following two questions: (1) which specific model of regulation could be appropriate to discipline peculiarities derived from the new kind of research on technoscientific subjects, and (2) which principles can model specific regulations.

All this is made somewhat complicated by the fact that with nanotechnology, the issue is one of 'futuristic' risk assessment issues ('anticipatory regulation', i.e. of legislating for an emerging and as yet poorly understood technology). This requires one 'to select or construct a particular future from innumerable alternatives, a process which will hopefully provide enough information to base a decision on' (Kaiser et al. 2009). Thus the seminal 2004 Royal Society Report segments the field into specific futures for which ethics can prepare – 'specified uncertainties' – and then matches such futures with existing institutions that presumably can 'cope with them'. Nanorisk decision-makers will be in the situation of 'decision-making with incomplete knowledge of outcomes' (as well as their associative probabilities) (Allhoff et al. 2010).

Although this is surely the problem with any new technology, Dupuy and Grinbaum (2006) argue that nanotechnology is more problematic than most: 'one serious deficiency, which hamstrings the notion of precaution, is that it does not properly gauge the type of uncertainty with which we are confronted at present', and so they argue for a more extreme approach, that of 'ethics beyond prudence'. By this, they appear to mean that we need new ideas of what prudence might mean, as the precautionary principle is insufficient. As Ferarri (2010) notes, we have an 'epistemically new situation characterised by uncertainty, ignorance, and ambiguity'.

This could imply new and much stricter regulation. Or it might perhaps lead to more 'open' regulation, involving a greater spread of opinion and consultation.

4.2 Nanoregulation in the EU

> We are still in the initial phases of (policy) development. There are not, so far, any internationally agreed definitions relating to the technology.... Recent proposals for renewing regulation on food additives have made this the first piece of regulation to include explicit reference to nanotechnology. (von Schomberg 2010)

Given the novelty of nanotechnology, is there an opportunity, as Faunce (2007) suggests, to shape policy in Europe to reflect a 'greater balance between private and public goods in two areas of primary concern to human well-being: medicine and biosecurity'?

With the 2004 Lisbon Strategy, Europe stated its aim of being 'the most competitive and dynamic knowledge-based economy in the world by 2010' (Lisbon Agenda 2010). A focus on nanotechnologies was seen as a key part of this strategy. During the formation of the 2004 European Nano-Electronics Initiative Advisory, European technology companies such as Philips, Nokia, Ericsson and AMD decided that if Europe wanted to lead the world, it would need to invest at least 6 billion euros per year to switch from micro- to nanoscale electronics. A public-private partnership charged to come up with and implement a European nano-electronics research agenda first met in 2004. Its goals included supporting research and investment in nano, speeding up innovation and productivity, the facilitation and acceleration of market penetration of new technologies, aligning research/technology with European policies and regulatory frameworks and increasing public awareness, understanding and acceptance of nanotechnologies (Adams and Williams 2006). A later group, the European Network of Excellence program (Nano2Life) evolved from 2004 to 2008, and representing 200 scientists, 23 research organisations and 12 countries, joined with industrial partners to identify regional centres, disciplines and expertise available for collaboration. Goals have included developing joint research projects on four major technical platforms: functionalisation, handling, detection and integration of nanodevices.

Given such economic impetus (even, arguably, economic determinism), what governance exists, and what ethical advisory bodies function in the EU?

4.2.1 EU Nanotechnology Advisory Groups: EGE, HLEG, ETAG and EGAIS

The history of EU nanoregulation should be placed in the context of biotechnology regulation, which dates back to the 1990s, with the establishment in 2001 of a European Commission Group on Ethics in Science and Technology (EGE). The group was created to advise the commission on how to exercise its powers as regards the ethical aspects of biotechnology. It noted in its general report for the 2005–2010 period examples of embedded ethics in policy such as the Science and Society Action Plan (2001), the Action Plan for Life Sciences and Technology (2007), as well as legislative activities including directives governing clinical trials, patents, data protection, the use of animals in experimentation and the EU Chemical, Biological, Radiological or Nuclear (CBRN) Action Plan.

Unsurprisingly, the advisory groups have advocated the precautionary principle. The EGE's 2006 Opinion 21 paper underlined 'the vital importance of addressing concern for safety with respect to ... nanotechnology in general'. It advocated, in regard to nanomedicine, the need to establish measures to verify the safety of nano-medical products and issues of military usage of nanotechnology, enhancement, economic equity and animal testing (familiar issues, as identified in Chap. 1) (Martinho da Silva 2009). The EGE group has been referred to as 'largely unknown', however, according to a public (admittedly global rather than EU) survey (European Commission 2009).

The High-Level Expert Group (HLEG) on Key Enabling Technologies was convened by the EC in 2010 to develop possible policy measures to promote the industrial take-up of key technologies such as nano by EU industries. The HLEG's SIG subgroup (special interest group, nanotechnology) offered a 2004 report, advocating both moral pluralism and the need for ethics to be an intrinsic part for technological advances (Staman 2004).[11] Another EU advisory group is ETAG, the European Technology Assessment Group, which runs projects on the potential environmental, health and safety risks of engineered nanomaterials (such as their project on chemical risk in 2006 and on human enhancement in 2008–2009). The EGAIS (Ethical Governance of Emerging Technologies) project, funded by the 7th Framework Programme (*Science in Society*) has as its mission 'to overcome the existent limitations of the current approaches to ethical governance in projects with technical development'.[12] The European Commission's Directorate-General and Services involved in nanotechnology number over a dozen, including 7 agencies for risk evaluation.[13]

4.2.2   EU Priorities in Nanogovernance

The above outlines how the EU is aware of the need for nanoethics to be part of technology development. Are there any particular directions for EU concerns within that broad area? EU awareness of nanotechnology issues began in earnest in 2002–2003, after discussion of the potential risks by the European Parliament's Green Party (Oud 2007) and the publication of the 2001 National Nanotechnology Initiative (NNI) in the USA. In terms of the EU's 'framework' structures for funding research and technological development, Framework Programme 6 (FP6) (2002–2006) indicated that nanotechnology had become a research priority, possibly stimulated by a need to compete. FP6 was introduced with policy objectives to enhance innovation and to 'change the European research landscape through the introduction of the integrated European Research Area (ERA), and create sustainable growth, increased employment and greater social cohesion' (Pandza et al. 2011).[14]

---

[11] For analysis of the SIG group, see de S Cameron (2007).

[12] EU ethics advising has seen (according to Cordis, the EU's information repository) an increased level of funding allocated in terms of the EU's 7th Framework Programmes (2007–2011).

[13] The full list can be seen at: http://ec.europa.eu/nanotechnology/links_en.html

[14] The article has a useful table outlining the framework programs and the focus on nanotechnology in each.

The first expert recommendations for the EC suggested a special emphasis on developing a nomenclature for nanoparticles, on assigning a new Chemical Abstract Service (CAS) registry number to engineered nanoparticles and on grouping and classifying nanomaterials with respect to categories of risk, toxicity and proliferation (Grunwald 2012).

The three EC Scientific Committees set up in 2004 for nanogovernance were: SCCS (the Scientific Committee on Consumer Safety) that looks at nanomaterials in consumer products; SCER (the Scientific Committee on Health and Environmental Risks) that looks at nanotechnology in food, as well as medical and environmental issues; and SCENHIR (the European Commission's Scientific Committee on Emerging and Newly Identified Health Risks) that looks at methodologies for risk issues of new technology such as nanotechnology.

The EC's Action Plan on Nanotechnology (2005), as well as considering possible adverse effects on health and the environment, also highlighted the ethical issues concerning nanotechnology's potential to contribute towards Millennium Development Goals such as the eradication of poverty and disease. As well as discussing public participation and education, the plan encouraged internationally cooperative work on nanoregulation. In 2007, the European Commission accepted the first implementation report (2005–2007) of the Action Plan. The second implementation report was adopted in 2009, with the statement that 'efforts to address societal and safety concerns must be continued to ensure the safe and sustainable development of nanotechnology' (Commission of the European Communities 2009).

In 2006 NGOs entered the social debate on nanotechnology (in Europe). Most NGOs focus on threats to health and the environment, issues of controllability and power, as well as questions of access and equity. Many of these organisations support a strong precautionary principle and conclude with proposals for a moratorium on the use of nanomaterials in products, particularly in food and cosmetics.

4.2.3 *EU Achievements in Nanoregulation:* EU-CoC *and* REACH

The brief discussion above shows how, although there are many forums for discussing nanotechnology issues, little is happening in terms of concrete policy. What has the EU actually achieved? Two significant regulatory attempts have been the EU Code (EU-CoC) and REACH.

In 2008–2009, the European Commission created 'EU-CoC', a Code of Conduct for Responsible Nanosciences and Nanotechnologies Research, intended to 'ensure, safe, ethical and sustainable nanosciences and nanotechnologies research in the European Union' (Commission of the European Communities 2008). The code guidelines were intended for the use in national strategic nanotechnology planning and funding distribution and to create greater public accountability.

The Code's seven principles – 'meaning, sustainability, precaution, inclusiveness, excellence, innovation, and accountability' – were generally accepted, although 'accountability' was felt to be too strong and should be replaced with 'responsibility' (Commission of the European Communities 2008). There has been some praise of the code:

> The recently released Code...developed by the European Commission is one code that is voluntary, but which has originated in a political sphere and which demands a higher level of accountability.... the code seeks to intervene at an earlier stage in the development cycle of nanotechnologies, embedding principles of responsibility at the research stage. (Davies et al. 2009)

However, it has also been criticised. The European Project *NanoCode*, a 2-year multi-stakeholder dialogue providing input to EU-CoC for Responsible Nanosciences & Nanotechnologies Research, commenced in January 2010, with the aims of monitoring stakeholder input and suggesting revision to this code of conduct. The issue of the 'lack of teeth' in voluntary codes such as this one is demonstrated by the fact that only 21 % of respondents' organisations had actually adopted the code. Suggested improvements have ranged from the development of tailor-made codes of conduct for specific nano-companies, 'naming and blaming' in the case of noncompliance, linking compliance to public funding and the incorporation of the code into the EC Research Framework as a guideline. There was a further issue of whether the code should be promoted as the 'one and only' or allowed to coexist with the institutional guidelines of each respondent's organisation. NanoCode's (2011) report on stakeholder attitudes towards the code concluded that 'awareness of the code was limited to a community of selected key experts', and was not embedded in the everyday life of the large majority of researchers in Europe. The report also concluded that few governments seemed able to communicate their principles to stakeholders effectively.

The second achievement of EU policy has been the 2006 EC establishment of REACH (the **R**egistration, **E**valuation, **A**uthorisation and Restriction of **C**hemical Substances), a regulatory framework addressing the following aspects of nanotechnologies: classification, terminology and nomenclature; metrology and instrumentation, including specifications for reference materials; test methodologies; modelling and simulation; science-based health, safety and environmental practices; and nanotechnology products and processes (European Committee for Standardization 2014).[15]

The aim of REACH is:

> to improve the protection of human health and the environment through the better and earlier identification of the intrinsic properties of chemical substances...The REACH Regulation places greater responsibility on industry to manage the risks from chemicals and to provide safety information on the substances. Manufacturers and importers are required

---

[15] See also 2.1.1 at http://www.ecostandard.org/downloads_a/cen-overview-std-nanotech-sept07. pdf. This website has an analysis of national standards bodies in the EU, UK, North America, Japan, China and Korea.

to gather information on the properties of their chemical substances, which will allow their safe handling, and to register the information in a central database ....One of the main reasons for developing and adopting the REACH Regulation was that a large number of substances have been manufactured and placed on the market in Europe for many years, sometimes in very high amounts, and yet there is insufficient information on the hazards that they pose to human health and the environment. There is a need to fill these information gaps to ensure that industry is able to assess hazards and risks of the substances, and to identify and implement the risk management measures to protect humans and the environment. (REACHCentrum 2008–2011)

In 2008, the EC Commission (Regulatory Aspects of Nanomaterials) concluded that existing EU regulatory frameworks covered in principle the potential health, safety and environmental risks related to nanomaterials and stressed that the protection of these areas would be enhanced by improving the implementation of current legislation instead. In answer to this communication, the European Parliament declared the EC's statement to be misleading and 'one dimensional' (Widmer and Knebel 2009). In 2010–2011, the European Parliament, concerned with the regulation of nanomaterials under REACH, again requested further study and responses to nanorisk. The concern was that it did not go far enough to deal with the dangers of nanochemicals, given that they operate at such a small scale.

This was the impetus behind the September 2011 German Advisory Council on the Environment's report *Precautionary Strategies for Managing Nanomaterials.* The report offers a number of suggestions and recommendations, including 'extensive changes in chemicals legislation (REACH)', arguing that 'nanomaterials should be consistently treated as if they were substances in their own right and registered with dossiers of their own, as well as that 'authorisation should be based more closely on the precautionary principle.'

In early February, 2012, the Center for International Environmental Law (CIEL) released an assessment of the status of regulating nanomaterials under REACH. The '*Just out of REACH*' CIEL report recognizes four areas where REACH is 'not living up to expectations for nanomaterials' and offers options for modifying REACH 'to fill the problematic knowledge gap on nanomaterials'. The report proposes developing a stand-alone regulation, 'carefully aligned with the chemical rules, but specifically tailored to nanomaterials, a nano "patch" that closes these inherent loopholes' (The Center for International Environmental Law 2012). Such a regulation would establish clear, legally binding provisions for nanomaterials and create a transparent and predictable legal environment for the safe production and use of nanomaterials in the EU.

The 2012 Communication on the Second Regulatory Review on Nanomaterials describes the commission's plans to improve EU law and its application to ensure their safe use, concluding that:

nanomaterials are similar to normal chemicals/substances in that some may be toxic and some may not. Possible risks are related to specific nanomaterials and specific uses. Therefore, nanomaterials require a risk assessment, which should be performed on a case-by-case basis...Current risk assessment methods are applicable, even if work on particular aspects of risk assessment is still required. (European Commission 2012)

These conclusions have been criticised for failing to implement the precautionary principle sufficiently. This is becoming a global problem:

> Analysts and regulatory reviews have called for increased international cooperation, standard-setting, and capacity-building. Some point to the growing internationalization of nanotechnology, and the burgeoning international trade in nanomaterials and products made with nanotechnology which are likely to create a greater need for international harmonization of regulations. (Falkner and Jaspers 2012)

Given EU-CoC and REACH, does more need to be done in terms of regulation and policy development? With a lack of definitive knowledge about the effects of nanotoxicity, the issue of how far the precautionary principle must be applied is still being debated. There are still issues about whether a 'truly' precautionary approach is in fact being followed in the EU at the moment (Strand and Kjølberg 2011).

However, it should also be noted that the one EU (2010-) ongoing public consultation project, 'SNAP', has thus far concluded that there was a general perception of nanotechnology benefits, particularly in communications/computing, energy, aerospace, construction, sustainable chemistry, security and the environment. Applications for medicine, agriculture, food and household items were regarded with more scepticism; yet there was a 'good or very good perception of EU governance related to nanotechnologies in terms of stakeholder consultation and setting research priorities', while respondents did 'not expect more actions in the fields of ethical, legal and social aspects of nanotechnology, or in terms of ensuring ethical reviews of EU nano R&D projects' (European Commission 2010).

4.3 Nanoregulation in China

China is arguably further behind on the nanotechnology governance challenges – such as addressing low public awareness, developing a robust risk research strategy and implementing an effective oversight system – compared with that encountered in the West. According to Tang et al. (2011), the initial focus for China was on creating regulatory support for significant research in technology commercialisation and economic growth. However, a 2004 conference on biological as well as environmental nanoeffects and the launch of a program on the toxicological effects of nanotechnology suggested that a nanosafety debate has begun in China, so that, as Qi (2008) has argued, nanosafety research has become 'an integral part of nanotechnology research'.

Now that markets are global, how will the public outside of China trust Chinese-made (nano)products in the wake of scandals involving tainted pet food, toothpaste, children's toys and drugs? Michelson (2008) asks:

> ...can China find an effective way to move up the value chain and transition from manufacturing cheaper, low-end products to more expensive, high-end, nano-engineered goods? ....
> In the wake of more immediate environmental and public health concerns currently affecting China – from managing pollution, to depleted fisheries, to lack of access to healthcare – how will disruptive challenges posed by nanotechnology exacerbate or enhance these

existing problems? Will China have the resources and the luxury of proactively addressing environmental and health risks posed by nanotechnology in addition to these ongoing challenges?[16]

The Chinese national program of nanotechnology development and regulation dates from 1987, to the National High Technology Plan (the 863 Plan, as it is called), which supported ultrafine particles research. In 1990, the State Science and Technology Commission [the predecessor of the current Ministry of Science and Technology (MOST)] approved the 'Climbing Up' project, in which nanomaterials research was emphasized. In the early 1990s, several Chinese academic research organisations joined together to accelerate research efforts in China in nanomaterials science, while MOST's 973 Program (1999) was aimed at supporting basic research on nanomaterials and nanostructures (e.g. nanotubes).

The 973 Program's significance lies in its emphasis on the standardisation of procedures along with assessment and test protocols, which form the basic framework for regulatory considerations of nanomaterials.[17] It comes under the review of the National Steering Committee for Nanoscience and Nanotechnology (NSCNN), established in 2000 to coordinate nationwide efforts on nanotechnology R&D. The committee is composed of 21 scientists from universities, institutes and industry as well as 7 administrators from government agencies.

In 2002, the 863 Plan established a series of projects that support nanomaterials and nanodevice applications. Since then, nanotechnology has been further recognized as a high-priority area by the Chinese government (Jia et al. 2011). In China's National Medium- and Long-term Science and Technology Development Plan (2006–2020), nanotechnology is identified as one of four priority mission areas/key frontier technologies over the next 15 years (Suttmeier et al. 2006). Appelbaum and Parker (2007) conclude that China is closing the nanogap that once existed between itself and the USA, Europe and Japan.

Presently, there are more than 30 research organisations in China that have initiated research activities studying the toxicological and environmental effects of nanomaterials and nanoparticles (Zhao et al. 2008), while 120 organisations are conducting general research into nanoscience and nanotechnology (Decker and Li 2009). China has published 15 nanotechnology standards since the establishment of its Committee of National Nanotechnology Standards (2005), dealing mainly with the risk assessment of nanoparticles.

China's regulatory regime for the management of nanotechnology chemicals will likely manage risks comparable to those identified under the EU's REACH regulatory framework, and in 2009 the Chinese government revised the chemical substance rules in order to incorporate risk assessment, risk management and data

---

[16] Also see Michelson (2007).

[17] The LSE/NUS (Lee Kuan Yew School of Public Policy at the National University of Singapore and the London School of Economics and Political Science) joint project, 'Nanotechnology Oversight in Asia and Europe' (2009–2010), examines emerging nanotechnology policies in China, India, Taiwan and the European Union. Within this project, Jarvis and Richmond (2010) argue for the significance to China's nanodevelopment of the 973 Program.

submission requirements similar to REACH (this regulation being colloquially known as 'China REACH') (Chemical Inspection & Regulatory Service 2014).

Of the published Chinese nanotechnology standards, 17 are voluntary, but as Jarvis and Richmond (2011) note, the regulatory environment is complex, and the terms voluntary' and 'mandatory' tend to be interpreted in terms of market access (as greater adherence to standardisation opens markets). This commercial agenda seems positive, but as China's S&T discourse is tied to the national political agenda, the emphasis on progress rather than caution suggests a differing emphasis to that in the EU (to a degree – economic drivers, however much countries euphemise them, – remain compelling). Thus:

> The emergence of an effective regulatory regime able to manage nanoscience-based-risks in China (is) problematic. The relatively closed nature of the nanoscience community and an absence of outreach or public engagement creates regulatory modalities that might better be characterised as…closed governance regimes. (Jarvis and Richmond 2011)

China has run two major 5-year projects: 'The Toxicological Effects of Carbon Nanomaterials' (2004–2008) and 'The Environmental Activity and Health Impact of Ambient Superfine Particles' (2006–2010). The China Nanosafety Lab, which examines environmental health and toxicology matters, is linked into a larger network of research centres for nanosafety. From its inception in 2008 with the establishment of the Laboratory for Biomedical Effects of Nanomaterials and Nanosafety, it has worked with the Research Centre for Cancer Nanotechnology (at the Tianjin Cancer Hospital), the Lab for the Bio-environmental Effects of Nanomaterials and Nanosafety (established by the Institute of High-Energy Physics or IHEP) and the NCNST or National Centre for Nanoscience and Technology and the nano-biological research group at IHEP.

The Chinese Ministry of Science and Technology (MOST) is generally responsible for the planning and execution of national S&T programs. It is also responsible for drafting rules, regulations and laws, as well as having responsibility for policy implementation. MOST is supported by the Chinese Academy of Science and Technology for Development (CASTED), an institution that undertakes strategic research to provide macro-level advice and assistance for designing S&T plans. Initially, emphasis was placed solely on scientific research in the nanotechnology field; however, new research teams such as the Institute of Policy and Management (IPM), part of the Chinese Academy of Sciences, have now been formed.[18]

CASTED contains ISTS, the Institute of Science, Technology and Society, which conducts several large-scale surveys on matters such as food safety risks and the public image of scientists. One of its three main areas of interest is the social environment of innovation.[19] CASTED has called for S&T ethics courses in universities on the basis that education is the main channel for building a culture of S&T values. Tsinghua University, for example, has an engineering ethics course.

---

[18] Professor Zhenzhen Li, Bilat-Silk (Bilateral Support for the International Linkage with China, FP7, 222800) roundtable on 9 November, 2011 at CASTED, Beijing, China (own notes of discussion).

[19] See the CASTED website at http://www.casted.org.cn/en/web.php?ChannelID=67

Several Chinese universities, including Dalian Technology University, Beijing University, Renmin University, Hunan and South-East, have research centres based around S&T ethics.

In conclusion, China, while having always had robust structures for risk governance, appears to be moving from a narrative of economic benefit towards an awareness of potential issues, both those relating to risk and to the more general 'NELSI' debate. China's similar-to-REACH regulation suggests it is in much the same position in policy terms as the EU; the difference however is marked in that there is no structure of NGO bodies to comment on deficiencies in REACH, as in the EU, or to survey the implementation of REACH in China, nor is there the level of civil society involvement, as in the EU. In terms of a code analogous to that of EU-CoC, Chinese S&T general guidelines (set out by MOST) have applied to nanoresearch for decades; a committee on science ethics examined scientific misconduct in 2010–2011 and has the ethics of emerging technologies (GM and nanotechnology) as its next focus, with a code predicted in 2013.

4.4   Global Nanopolicy

International ethical guidelines need to be formulated, not only in general as opposed to particularistic terms but in such general terms as would make sense and meaning to variously and differently situated and circumstanced human communities, groups, and assemblages....What a good international ethical guideline requires in its formulation and expression is a balancing of different but not necessarily conflicting points of view and perspectives – the underlying ethical imperative alone remaining constant. (Tangwa 2004)

A Chatham House briefing paper in 2009 identified several nanotechnology challenges including the pace of change, the uncertainty of commercialisation paths and the suitability of regulatory frameworks and of resourcing. The recommendations included:

- Closing the 'knowledge gap' through greater funding for research into nanorisk, including greater research funding coordination and better data sharing (a problem given corporate confidentiality issues)
- Improving market registers with a view to global coordination and to 'gaining a comprehensive overview of the commercial use of nanomaterials'
- Increasing work on establishing a firm scientific base for risk assessment, through greater resourcing of the OECD global nanotechnology working parties
- Giving greater consideration to global governance challenges, with greater representation of developing countries in global regulatory cooperation (Falkner et al. 2009)

The above emphasis on 'sharing', global risk assessment and global governance challenges indicates an awareness of the need for global risk regulation.

Several of UNESCO's member states have mandated that the organisation will set universal ethical benchmarks covering issues raised by the rapid development of

S&T. Will there be a similar declaration to the universal one on bioethics and human rights for nanotechnology or even a 'Nanotechnology Protocol' similar to the (binding) Cartagena Protocol on Biological Diversity?[20] The latter aims to ensure the safe handling, transport and use of living modified organisms. It is the result of concerns about how modern biotechnology may have adverse effects on biological diversity, also taking into account potential risks to human health. Is a global framework convention advisable for nanotechnology? (Abbott et al. 2008). The ETC (2006) (a technology-tracking civic action group) have argued that an 'International Convention for the Evaluation of New Technologies' at the United Nations is needed, namely, 'an intergovernmental framework that would allow for the monitoring and evaluation of new technologies as they evolve from initial scientific discovery to possible commercialization'.

Any formal international regulatory agreement for nanotechnology will face many obstacles and challenges, so 'if it is likely that any formal international regulatory agreement is many years in the future…it may be worthwhile to consider other, less formal alternatives to binding treaties at the international level' (Marchant and Sylvester 2006).[21] Some options along this line include the code of conduct already developed in the EU. A UNESCO paper on ethics and scientific development in Malaysia, noting the question of whether a Universal Oath for Scientists was necessary, suggested that differences between, say, the Hippocratic Oath, or the Seventeen Rules of the Enjuin (an alternative Japanese code of medical ethics), or the Islamic Oath for the physician, are not as marked as are often thought (Mohd Nor 2007).[22]

The 1996 EU Oviedo Charter (the 'Convention for the Protection of Human Rights and Dignity of the Human Being with regard to the Application of Biology and Medicine: Convention on Human Rights and Biomedicine') was drawn up in response to the argument that advances in biomedicine were moving at such a pace that the laws in the various EU member states were not able to keep pace with the developments. The concern was that the rapid pace of developments and the fragmentation of approaches among member states allowed for the possibility of 'havens' to emerge for research, in which scientists could exploit a lack of regulation so as to evade legal restrictions in their own countries (Petroni 2006).[23]

The proper implementation of regulation requires for a transnational dialogue and information sharing, like The International Dialogue on Responsible Research and Development of Nanotechnology (which included representatives of 49 counties) (Tomellini and Giordani 2008), as well as international standards and export controls. Examples include the awareness-raising workshops on nanotechnology that were held in Beijing, Abidjan, Lodz, Kingston and Alexandria. There have also been International dialogues such as that on the Responsible Research and

---

[20] See this protocol at http://bch.cbd.int/protocol/

[21] See also Reynolds (2003).

[22] The Islamic Code of Medical Professional Ethics, based on Qur'anic ethics, although giving clear advice about respecting patients, fails to mention autonomy as a special premise.

[23] Not all EU countries have signed the Oviedo charter.

Development of Nanotechnology (established by The European Commission in 2004) and the Meridian Institute's Global Dialogue on Nanotechnology and the poor, which has resulted in a paper, a news service and two workshops (India and Brazil, 2006 and 2007).[24]

A November 2011 project on research methods for managing the risk of engineered nanoparticles and engineered nanomaterials, the MARINA project, funded under FP 7 of the EC, will run for 4 years and involve the collaboration of 47 different scientific and industrial partners, including China. There are also individual joint country agreements, such as the 2002 EU/China cooperation agreement in the field of material sciences. This agreement facilitates the participation of Chinese research organisations, including companies, in European research projects with Chinese funding and vice versa. In addition, the EU and China have a joint agreement to exchange data relating to safety testing in order to boost research into the effect of nanotechnology products on consumer safety.

The European Standards Committee (CEN/TC352) Nanotechnologies was set up with the aim of providing an international standard. To date, the following international bodies have attempted some form of global supervision or regulation of nanotechnology:

1. The International Council on Risk Governance (IRGC)
2. The International Council on Nanotechnology (ICON) (Kulinowski 2007)
3. The Globally Harmonised Scheme for Classification and Labelling of Substances (GHS)
4. The International Standards Organization (ISO)
5. The Commission on the Ethics of Scientific Knowledge and Technology (UNESCO/COMEST)
6. The OECD Working Party on Nanotechnology (WPN), with a focus on governance
7. The OECD Working Party on Manufactured Nanomaterials (WPMN), an international forum for the further development of test guidelines and strategies

It has also been argued that the WTO (World Trade Organisation) has a role to play in the global regulation of nanotechnology (in that the WTO has a mandate to review any regulations with an impact on trade); yet it does not have the mandate or manpower to carry out any harmonisation of differing agendas (van Caster 2007).

There are other bodies that could be listed; however, these are organisations such as Committee E56 on Nanotechnology, which rather vaguely lists one of its activities as 'appropriate global liaison relationships with activities related to nanotechnology'. Set up in 2005 under the auspices of the American Society for Testing and Materials, it may have global reach, but it is debatable whether it is a global society in the sense of having international partners in research.[25] Another that fails to make the list above is the International Nanotechnology and Society Network that con-

---

[24] See the Meridian Institute's website at http://www.merid.org/en/Content/Projects/Global_Dialogue_on_Nanotechnology_and_the_Poor.aspx?view=cal

[25] See the ASTM International committee website at http://www.astm.org/COMMITTEE/E56.htm

sists of researchers exploring the connections between society and the possible upcoming changes provided by nanotechnology research. The members represent 37 institutions from 11 countries – but the last meeting referred to on the website is one from 2006.[26]

What do these seven bodies contribute to the global regulation of nanotechnology or to global collaboration on the regulatory consequences of this new technology?

Global or Not-So Global? The *IRG*, established in 2003 on the initiative of the Swiss government, has members from 13 different countries, one of which is China. Its main backers are apparently the Swiss, the USA and Chinese Governments, Swiss Re, Allianz, EON Energie, ATEL and the Swiss Federal Institute of Technology (not a particularly globally inclusive body but arguably one with global reach).[27] The IRG's recommendations include an improved knowledge base, the need to take societal concerns into consideration in risk evaluation, strengthened risk management procedures and processes and stakeholder participation (Renn and Roco 2006).

Similarly, *ICON* has the mission 'to develop and communicate information regarding potential environmental and health risks of nanotechnology, thereby fostering risk reduction while maximizing societal benefit'; however, it is local rather than global. Created between 2004 and 2007 by the National Science Foundation Center for Biological and Environmental Nanotechnology (CBEN) at Rice University in Houston, Texas, ICON/CBEN lists as current projects in 2011 the 'GoodNanoGuide', a collaborative platform for researchers to pool knowledge on how best to handle nanomaterials, workshops on classifying nanomaterials and developing predictive models for their interaction with living systems and a survey-based best practice project. ICON held an international workshop in March 2011 on policy, business and legal nanoissues; however, the majority of panellists listed were American, and no representatives from Asia were included (Marchant 2011). It is therefore difficult to assess the extent to which it might be called 'international'.

Global Standardisation: The GHS and the ISO These rules for the classification, labelling and safety data sheets of chemicals at national, regional and worldwide level have been promoted by the UN – but the GHS is nonlegally binding in the member countries of United Nations. Thus, many countries and regions have published their own regulations or standards to implement the GHS (Chemical Inspection & Regulatory Service 2013).

The EU adopted the GHS in 2008; it has been applied since January 20, 2009. China implemented it in 2010. Japan, Taiwan, New Zealand and Korea have implemented

---

[26] See the International Nanotechnology and Society Network website, under 'Meetings', at www. nanoandsociety.com

[27] See the IRGC website at http://www.irgc.org/about/. The Chinese member is Professor Hou Yunde, Director, State Centre for Viro-Biotech Engineering and State Key Laboratory for Molecular Virology and Engineering, Beijing, People's Republic of China.

GHS in the last 3 years, while South Africa's 2007 National Standard essentially followed the GHS template but needs expansion for full GHS regulation.[28]

The various ISO standards largely provide methodological tools for industry to evaluate risk. However, it should be noted that the ISO has a member list of 162 countries, including China, which has been involved in 707 standards more than any other country listed.[29] China hosted ISO/TCC 229 in 2008 in Shanghai.

Genuinely Global? Issues of nanotechnology and ethics were first explored during the Third Session of *COMEST* (The Commission on the Ethics of Scientific knowledge and Technology UNESCO/COMEST) in Rio de Janeiro in December 2003, which was set up under the auspices of UNESCO. UNESCO will likely play a leading role in the area of nanotechnology, ensuring that the perceptions and interests of developing countries are duly considered while insisting on the importance of tackling this subject creatively from a global perspective (UNESCO 2006a, b).

In 2005, a group of experts was established in order to assist COMEST in drafting a potential policy document for ethics and nanotechnology. The group included representatives from South Korea, Canada, New Zealand, Japan, the Netherlands, China, Germany and Brazil. This group reported results from two meetings in 2005 and provided input into COMEST's draft, providing the basis for a consultation process which allowed the development of a set of recommendations for the 34th UNESCO General Conference.

At the COMEST Extraordinary Session held in Paris in November 2008, COMEST members expressed their concern about the lack of visibility of this policy document and the failure to take steps towards its implementation, given the serious ethical issues emphasized by past work. COMEST members emphasised the potential toxicity of nanoengineered particles and the development of military applications, as well as the need for informed public debate on the ethical issues raised by nanotechnologies. The COMEST Working Group on the Ethics of Nanotechnology meets regularly to update this policy document and its implementation strategies as well as to make the policies adopted by various countries more visible. The 2011 meeting noted the following:

- That nanotechnology is developing at such a rate that its short and long-term impact is sometimes difficult to identify.
- That science and technology are being driven by the wrong kind of interests, in particular military interests.
- That in third world countries progress (according to some) is unacceptably slow.
- Risk management of nanomaterials and their use in consumer products continues to give cause for concern.[30]

---

[28] Details of the Globally Harmonized System can be found at their website. http://www.sigmaaldrich.com/safety-center/globally-harmonized.html

[29] A members' list can be found at http://www.iso.org/iso/about/iso_members.htm

[30] World Commission on the Ethics of Scientific Knowledge and Technology (COMEST) Working Group on the Ethics of Nanotechnology (Brussels, 27–28 April 2011), at http://unesdoc.unesco.org/images/0021/002197/219777E.pdf

4.4.1 The OECD (WPM and WPMN)

The OECD, through its Chemicals Committee and the mutual recognition arrangement (the Cooperative Chemical Assessment Program', that allows for international harmonisation of regulation on new chemicals), has a role to play in the regulation of nanotechnology. The OECD Working Party on Nanotechnology (WPN) has been examining how nanotechnology can contribute to the sustainable provision of clean water through improved water filtration devices (Rickerby and Carbone 2008).

However, international harmonisation can vary across the 30 member states of the OECD (Visser 2007).

As Kurath notes, perhaps none of the governance measures, soft law and self-regulatory schemes currently operating relative to nanotechnology appears socially robust in all aspects. He does treat the OECD Working Party on Manufactured Nanomaterials (WPMN) positively though, as it has 'comprehensible standards', 'is sensitive to risk' and 'has a steering committee that may play a role in the political translation of results and in evaluation' (Kurath 2009).

The OECD (2011) report, *Nanosafety at the OECD: The First Five Years 2006–2010*, notes the following outcomes, among others:

- The launch in 2009 of the *OECD Database on Manufactured Nanomaterials to Inform and Analyse EHS Research Activities,* a global resource for research projects that address environmental, human health and safety (EHS) issues of manufactured nanomaterials.[31]
- The launch in 2007 of the Sponsorship Programme for the Testing of Manufactured Nanomaterials to provide information on the intrinsic properties of nanomaterials crucial in choosing or adapting existing risk evaluation and management strategies. WPMN has developed an IT collaborative platform called 'Communities of Practice' (CoP), which refers to groups of experts who convene to discuss technical issues related to testing being conducted under the Sponsorship Programme.
- OECD and UNITAR (United Nations Institute for Training and Research) jointly held Awareness-Raising Workshops on Nanotechnology.
- National surveys to examine various national voluntary reporting schemes and regulatory programs to assess the safety of manufactured nanomaterials. Surveys are underway among national bodies with the aim of updating trends relating to commercial activities and their regulatory oversight and collecting specific information on manufactured nanomaterials (such as types and volumes used).

Global regulation *specifically* focused on nanotechnology is, as it is at the local level, non-existent, while existing global regulations on nanotechnology risk covered by existing (mainly chemical) regulations such as REACH are being challenged by NGOs. Global cooperation still appears alive and well and may lead to more voluntary codes or perhaps place more pressure on governments to regulate. It

---

[31] See this database at www.oecd.org/env/nanosafety/database

certainly will lead to the dissemination of knowledge among international bodies and scientists and could make a greater amount of information available to the public. More significantly, it will reduce the 'closed' nature of the Chinese nanoscientific community, possibly creating greater public debate.

4.4.1.1 The Discussion Thus Far

In the chapters thus far, discussion has ranged over the background context to global ethics – offering social values and bioethics regulation and various attempts to improve global collaboration on standards. The differences noted so far between the EU and China suggest conclusions such as that:

- The difference in the level of public engagement is most salient.
- The discourses of precaution versus economic progress may suggest a major difference, yet the precautionary principle in the EU is not 'truly' being followed according to certain groups, for economic imperatives are global and regulatory mechanisms not always adequate or mandatory.
- Similar ethical concerns have been expressed, in that toxicity risk is paramount, particularly given previous scandals relating to food.
- The bioethics contexts reveal a social distinction between individualism and communitarianism, shown, for example, by bioethics laws in the EU that privilege individual human rights.
- There is a contrast between an 'open' system, where civil society engagement is widespread, and a 'closed' one, where decision-making is top-down and centralised (and usually less accountable).

The following table summarises some of the comparative discussion across the chapters thus far (Table 4.1).

Table 4.1 A comparison of nanoethics environments in the EU and China

	EU	China
Social values:	Autonomy, individual human rights	Harmony, community
Attitude towards S&T and scientists:	Public differentiation between technologies, i.e. differing risks for nuclear, nano, GM	Innovation is always good
		Harmony between *tian* (nature) and *ren* (human)
	Risk vs. benefit	Trust in government (although this may be starting to decrease due to food scandals)
	Post-Fukushima emphasis on confidence/or lack thereof in government, according to how it handles S&T crises	
		Concern with social responsibility of scientists
GM food as example	GM 'backlash' in many EU countries	GM products sold in China but must be labelled; some public distrust

(continued)

Table 4.1 (continued)

	EU	China
Which nanoissues are seen as significant?	Toxicity (human and environmental), enhancement, military use, privacy, distributive justice	Safety; science governance; global security an emerging issue, as is public knowledge deficit
Economic benefit focus	Better drugs, e.g. for cancer treatment	General economic driver
Consumer confidence in new products	EU 2013 mandatory labelling of cosmetics that contain nanoparticles	New equals good
		'Nano' a positive label
Institutes leading nanoresearch	EC *Action Plan on Nanotechnology 2005–2009* (ongoing implementation reports in 2007, 2009)	2000: National Steering Committee for Nanoscience and Nanotechnology established in China
Seminal reports that have triggered nanoethics concerns:	Greenpeace report, the Royal Society and Royal Academy (RS/RAE), report and insurance company Swiss Re's report (2003–2004)	
Role of NGOs in S&T advising:	Strong in terms of TA and policymaker liaison	NGOs small in size and relatively weak in organisational capacity, no policy-effecting channels
TA infrastructure	Varied across Europe	Established for about 10 years
Participatory Technology Assessment	Many European dialogue initiatives, plus individual (national) initiatives in Germany, the Netherlands, UK, etc.	Consensus conference 2008 on GM food – with 25 participants only
Global view and cooperation	On standards	On standards
		Increasing awareness of Western debates
International bodies – IRG, REACH, COMEST, OECD, etc.	EU representatives on all	China active on ISO and REACH
Codes?	EU Code for responsible nanoresearch	MOST S&T general guidelines applied to nanoresearch from beginning
		CAS Special Committee on science ethics examined scientific misconduct in 2010–11 and has ethics of emerging technologies as next focus (GM and nano)

Much of this information was taken from roundtable discussions in Beijing 9–11 November, 2011

References

Abbott KW, Marchant GE, Sylvester DJ (2008) A framework convention for nanotechnology? Environ Law Report 38:10507–10514

Adams W, Williams L (2006) Nanotechnology demystified. McGraw-Hill Professional Publishing, Blacklick

Allhoff F, Lin P, Moore D (2010) What is nanotechnology and why does it matter. Wiley-Blackwell, Chichester

Anderson A, Petersen A, Wilkinson C, Stuart A (2009) Nanotechnology, risk and communication. Palgrave Macmillan, Basingstoke

Appelbaum RP, Parker RA (2007) Innovation or imitation? China's bid to become a global leader in nanotechnology. http://www.cns.ucsb.edu/index.php?option=com_remository&Itemid=100 &func=select&id=3P. Accessed 30 Nov 2014

Arnall A (2003) Future technologies, Today's Choices. http://www.greenpeace.org.uk/MultimediaFiles/Live/FullReport/5886.pdf. Accessed 30 Nov 2014, 20 Dec 2014

ASTM International (1996–2004) Committee E56 on nanotechnology. http://www.astm.org/COMMITTEE/E56.htm. Accessed 30 Nov 2014

Avolainen KS, Pylkkänen L, Norppa H, Falck G, Lindberg H, Tuomi T, Vippola M, Alenius H, Hämeri K, Koivisto J, Brouwer D, Mark D, Bard D, Berges M, Jankowska E, Posniak M, Farmer P, Singh R, Krombach F, Bihari P, Kasper G, Seipenbusch M (2010) Nanotechnologies, engineered nanomaterials and occupational health and safety – a review. Saf Sci 48(8):957–963

BASF (2013) Dialogforum nano of BASF 2011|2012. Transparency in communication on nano-materials from the manufacturer to the consumer. http://www.dialogbasis.de/fileadmin/content_images/Home/Dialogforum_Nano_of_BASF_2011-2012_web_engl.pdf. Accessed 26 Nov 2014

Berger M (2012) Blowing hot air – how not to criticize nanotechnology. Nanowerk. http://www.nanowerk.com/spotlight/spotid=19060.php. Accessed 30 Nov 2014

Bowman DM, Hodge GA (2008) 'Governing' nanotechnology without government?' Sci Public Policy 35:475–487

Buzea C, Pacheco I, Robbie K (2007) Nanomaterials and nanoparticles: sources and toxicity. Biointerphases 2(4):MR17–MR71

CEN (2007) International standards in nanotechnology. http://www.ecostandard.org/downloads_a/cen-overview-std-nanotech-sept07.pdf. Accessed 30 Nov 2014

Chemical Inspection & Regulatory Service (2013) GHS in China, South Korea and Japan. http://www.cirs-reach.com/GHS_in_China_Korea_and_Japan.html. Accessed 30 Nov 2014

Chemical Inspection & Regulatory Service (2014) New chemical substance notation in China. http://www.cirs-reach.com/China_Chemical_Regulation/IECSC_China_REACH_China_New_Chemical_Registration.html. Accessed 30 Nov 2014

Cheng J, Xu Xin H, Cao X, Cheng S (2009) Risk management of nanotechnology. J Saf Manage 9(3):148–154

Chinese Academy of Science and Technology for Development (CASTED) (2008) Introduction of departments and institutes. http://www.casted.org.cn/en/web.php?ChannelID=67. Accessed 30 Nov 2014

Cobb M, Macoubrie J (2004) Public perceptions about nanotechnology. J Nanopart Res 6:395–405

Commission of the European Communities (2008) Commission recommendation of 07/02/2008 on a code of conduct for responsible nanosciences and nanotechnologies research. http://ec.europa.eu/research/science-society/document_library/pdf_06/nanocode-recommendation-pe0894c08424_en.pdf. Accessed 30 Nov 2014

Commission of the European Communities (2009) Nanosciences and nanotechnologies: an action plan for Europe 2005–2009. Second implementation report 2007–2009. http://eur-lex.europa.eu/LexUriServ/LexUriServ.do?uri=COM:2009:0607:FIN:EN:PDF. Accessed 30 Nov 2014

Convention on Biological Diversity (2014) The Cartagena Protocol on biosafety. http://bch.cbd. int/protoco/. Accessed 30 Nov 2014

Davies S, Macnaghten P, Kearnes M (eds) (2009) Reconfiguring responsibility. Deepening debate on nanotechnology. A research report from the DEEPEN project. http://www.geography.dur. ac.uk/Projects/Portals/88/Publications/Reconfiguring%20Responsibility%20September%20 2009.pdf. Accessed 30 Nov 2014

de S Cameron NM (2007) Ethics, policy, and the nanotechnology initiative: the transatlantic debate on 'Converging technologies'. In: de S Cameron, Mitchell EM (eds) Nanoscale, issues and perspectives for the nano century. Wiley, Hoboken, pp 279–294

Decker M, Li Z (2009) Dealing with nanoparticles: a comparison between Chinese and European approaches to nanotechnology. In: Ladikas M (ed) Embedding society in science & technology policy. European and Chinese perspectives. European Commission, Brussels, pp 91–123

Dupuy JP, Grinbaum A (2006) Living with uncertainty: towards the ongoing normative assessment of nanotechnology. In: Schummer J, Baird D (eds) Nanotechnology challenges – implications for philosophy, ethics and society. World Scientific Publishing, Singapore, pp 287–314

Elder A (2006) Tiny inhaled particles take easy route from nose to brain. http://www.sciencenews-den.com/2006/tinyinhaledparticlestakeeasyroutefromnosetobrain.shtml. Accessed 30 Nov 2014

ETC Communique (2006) The international convention for the evaluation of new technologies. http://www.precaution.org/lib/08/icent.050601.htm. Accessed 30 Nov 2014

European Commission (2004) Nanotechnologies: a preliminary risk analysis on the basis of a workshop organized in Brussels on 1–2 March 2004 by the Health and Consumer Protection Directorate General of the European Commission. http://ec.europa.eu/health/ph_risk/docu-ments/ev_20040301_en.pdf. Accessed 30 Nov 2014

European Commission (2010) Report on the European Commission's public online consultation. Towards a strategic nanotechnology action plan (SNAP) 2010–2015. http://ec.europa.eu/ research/consultations/snap/report_en.pdf. Accessed 30 Nov 2014

European Commission (2012) Communication from the commission to the European Parliament, the Council and the European Economic and Social Committee second regulatory review on nanomaterials. http://eur-lex.europa.eu/legal-content/EN/TXT/HTML/?uri=CELEX:52012D C0572&from=EN. Accessed 30 Nov 2014

European Commission (2014) Policy issues. http://ec.europa.eu/nanotechnology/links_en.html. Accessed 15 Dec 2014

European Commission Directorate-General for Health & Consumers (2009) Scientific committee on emerging and newly identified health risks, SCENHIR. Risk assessment of products of nanotechnologies. http://ec.europa.eu/health/ph_risk/committees/04_scenihr/docs/ scenihr_o_023.pdf. Accessed 30 Nov 2014

European Committee for Standardization (2014) Fields of work. http://www.cen.eu/work/areas/ pages/default.aspx. Accessed 30 Nov 2014

Evans D (2007) Ethics, nanotechnology and health. In: Tenhave H (ed) Nanotechnologies, ethics, and politics. UNESCO, Paris, pp 125–154

Falkner R, Jaspers N (2012) Regulating nanotechnologies: risk, uncertainty and the global governance gap. http://static.squarespace.com/ static/538a0f32e4b0e9ab915750a1/t/538db0a1e4b0369e0b945df4/1401794721117/Falkner_ Jaspers_2012_Regulating_Nanotechnologies_final_ms.pdf. Accessed 30 Nov 2014

Falkner R, Breggin L, Jaspers N, Pendergrass J, Porter R (2009) Regulating nanomaterials: a trans-atlantic agenda. http://eprints.lse.ac.uk/25424/1/Regulating_nanomaterials_a_transatlantic_ agenda%28LSERO%29.pdf. Accessed 30 Nov 2014

Faunce TA (2007) Nanotechnology in global medicine and human biosecurity: private interests, policy dilemmas, and the calibration of public health law. Global Health Law, Ethics Policy 35(Winter):629–642

Fernholm A, Ingvald E, Sjostedt E (2008) Where gold glints blue: scientists on the nanorevolution. Vetenskapsradet, Stockholm

Ferrari A (2010) Developments in the debate on nanoethics: traditional approaches and the need for new kinds of analysis. NanoEthics 4:27–52

Fiedeler U, Nentwich M, Gressler S, Gazslo A, Simko M (2010) Voluntary approaches by industry in the field of nanomaterials. Nano-Trust Dossier 016en (December). http://epub.oeaw.ac.at/ita/nanotrust-dossiers/dossier016en.pdf. Accessed 30 Nov 2014

Gerloff K, Albrecht C, Boots AW, Forster I, Roel PF (2009) Cytotoxicity and oxidative DNA damage by nanoparticles in human intestinal Caco-2 cells. Nanotoxicology 3(4):355–364

German Advisory Council on the Environment (2011) Precautionary strategies for managing nanomaterials. http://www.umweltrat.de/SharedDocs/Downloads/EN/02_Special_Reports/2011_09_Precautionary_Strategies_for_managing_Nanomaterials_KFE.pdf?blob=publicationFile. Accessed 30 Nov 2014

Grunwald A (2012) Responsible nanobiotechnology: Philosophy and ethics. CRC Press, Boca Raton

Guerra G (2008) European regulatory issues in nanomedicine. NanoEthics 2:87–97

Hassan E, Sheehan J (2003) Scaling-up nanotechnology. OECD Observer No. 237. http://www.oecdobserver.org/news/fullstory.php/aid/1005/Scaling-up_nanotechnology.html. Accessed 30 Nov 2014

Holmes B (2004) Carbon "footfalls" harm fish. New Sci 182:11

Hunt G (2008) Nanotechnologies and society in Europe. In: Hunt G, Mehta M (eds) Nanotechnology. Risk, ethics and law. Earthscan, London, pp 92–104

Hussain S, Thomassen LCJ, Ferecatu I, Borot MC, Andreau K, Martens JA, Fleury J, Baeza-Squiban A, Marano F, Boland S (2010) Carbon black and titanium dioxide nanoparticles elicit distinct apoptotic pathways in bronchial epithelial cells. Part Fibre Toxicol 7(10). http://www.particleandfibretoxicology.com/content/7/1/10/. Accessed 30 Nov 2014

Illuminato I, Miller G (2010) Nanotechnology, climate and energy: over-heated promises and hot air. http://www.nanopodium.nl/CieMDN/content/Nanotechnology_Climate_and_energy.pdf. Accessed 30 Nov 2014

ISO (2014) ISO Members. http://www.iso.org/iso/about/iso_members.htm. Accessed 30 Nov 2014

Jarvis D, Richmond N (2010) Mapping emerging nanotechnology policies and regulations: the People's Republic of China. http://www.lse.ac.uk/internationalRelations/centresandunits/regulatingnanotechnologies/nanopdfs/China2010.pdf. Accessed 30 Nov 2014

Jarvis D, Richmond N (2011) Regulation and governance of nanotechnology in China: regulatory challenges and effectiveness. Eur J Law Technol 2(3). http://ejlt.org/article/view/94/155. Accessed 30 Nov 2014

Jia L, Zhao Y, Liang X (2011) Fast evolving nanotechnology and relevant programs and entities in China. Nanotoday 6(1):6–11

Kaiser M, Kurath M, Maasen S, Rehmann-Sutter C (eds) (2009) Governing future technologies: nanotechnology and the rise of an assessment regime. Springer, Dordrecht

Kjølberg K, Wickson F (2007) Social and ethical interactions with nano: mapping the early literature. NanoEthics 1:89–104

Kulinowski KM (2007) The International Council on Nanotechnology: a new model of engagement. In de S Cameron NM, Mitchell NE (eds) Nanoscale, issues and perspectives for the nano century. Wiley, Hoboken, pp 393–412

Kurath M (2009) Nanotechnology governance: accountability and democracy in new modes of regulation and deliberation. Sci Technol Innov Stud 5(2):87–110

Lisbon Agenda (2010) at http://www.euractiv.com/en/agenda2004/lisbon-agenda/article-117510. Accessed 30 Nov 2014

Makoni M (2010) Case study: South Africa uses nanotech against TB, 24 November 2010. http://www.scidev.net/en/health/nanotechnology-for-health/features/case-study-south-africa-uses-nanotech-against-tb-1.html. Accessed 30 Nov 2014

Marchant G (2011) Nanotech conference tackles big policy questions for the "small" science. ASU News. https://asunews.asu.edu/20110216_nanotechlawconference. Accessed 26 Nov 2014

Marchant G, Sylvester DJ (2006) Transnational models for regulation of nanotechnology. J Law Med Ethics 34(4 Winter):714–25

Martinho da Silva P (2009) Nanotechnologies for sustainable development. Available via European Commission. http://ec.europa.eu/nanotechnology/pdf/swedish-presidency-event/martinho_de_silva.pdf. Accessed 30 Nov 2014

Maynard AD (2007) Nanotechnologies: overview and issues. In: Opopol N, Luster MI, Simeonova PP (eds) Nanotechnology – toxicological issues and environmental safety. Springer, Dordrecht, pp 1–14

McCray WP (2009) Will small be beautiful: making policies for our nanotech future. In: Bijker WE, Bernard Carlson W, Pinch T (eds) Technology and society. Building our sociotechnical future. MIT Press, Cambridge, pp 323–368

McKibben B (2003) Enough: staying human in an engineered age. Henry Holt, New York

Meridian Institute (2014) Global dialogue on nanotechnology and the poor. http://www.merid.org/en/Content/Projects/Global_Dialogue_on_Nanotechnology_and_the_Poor.aspx. Accessed 30 Nov 2014

Michelson ES (2007) Nanotechnology policy: an analysis of transnational governance issues facing the United States and China. In: Blanpied WA, Gang Z (eds) Proceedings of the US–China Forum on Science and Technology Policy. George Mason University, Arlington, pp 345–358

Michelson ES (2008) Globalization at the nano frontier: the future of nanotechnology policy in the United States, China, and India. Technol Soc 30(3–4):405–10

Mnyusiwalla A, Daar AS, Singer P (2003) Mind the gap. Science and ethics in nanotechnology. Nanotechnology 14:R9–R13

Mohd Nor SN (2007) Philosophical and practical reflections of Malaysian science. In: UNESCO Asia-Pacific perspectives on the ethics of science and technology. http://unesdoc.unesco.org/images/0015/001550/155000e.pdf. Accessed 30 Nov 2014

Motavalli J (2010) Wanted: nano-cops. new haven independent. http://newhavenindependent.org/index.php/archives/entry/wanted_nano-cops/id_26623. Accessed 30 Nov 2014

Nanocode (2011) Stakeholder's attitudes towards the European code of conduct for nanosciences & nanotechnologies research synthesis report. http://www.phantomsnet.net/files/nanocode-consultation-synthesis-report.pdf. Accessed 30 Nov 2014

Nanoposts.com (2007) International standards in nanotechnology. http://www.ecostandard.org/downloads_a/cen-overview-std-nanotech-sept07.pdf. Accessed 1 Dec 2014

Nanowerk (2012) NGOs respond to European Commission's second regulatory review of nanomaterials. http://www.nanowerk.com/news2/newsid=27087.php. Accessed 30 Nov 2014

Nanowerk (2013) Nanotechnology policy making – mandatory tools. http://www.nanowerk.com/spotlight/spotid=29822.php. Accessed 30 Nov 2014

Oberdorster E (2004) Manufactured nanomaterials (Fullerenes, C60) induce oxidative stress in the brain of juvenile largemouth bass. Environ Health Perspect 112(10):1058–62

OECD (2009) OECD database on manufactured nanomaterials to inform and analyse EHS research activities. http://www.oecd.org/chemicalsafety/nanosafety/oecddatabaseonresearchintothesafetyofmanufacturednanomaterials.htm. Accessed 30 Nov 2014

OECD (2011) Nanosafety at the OECD: the first five years 2006–2010. http://www.oecd.org/science/safetyofmanufacturednanomaterials/47104296.pdf. Accessed 30 Nov 2014

Office Journal of the European Union (2009) Regulation(EC) No 1223/2009 of the European Parliament and of the Council of 30 November 2009 on cosmetic products. http://eur-ex.europa.eu/LexUriServ/LexUriServ.do?uri=OJ:L:2009:342:0059:0209:en:PDF. Accessed 30 Nov 2014

Oud M (2007) A European perspective. In: Hodge G, Bowman D, Ludlow K (eds) New global frontiers in regulation. The age of nanotechnology. Edward Elgar, Cheltenham, pp 265–86

Pandza K, Wilkins TA, Alfoldi EA (2011) Collaborative diversity in a nanotechnology innovation system: evidence from the EU Framework Programme. Technovation 31(9):476–489

Petroni AM (2006) Perspectives for freedom of choice in bioethics and health care in Europe. In: Engelhardt HT Jr (ed) Global bioethics: the collapse of consensus. M&M Scrivener Press, Salem, pp 238–270

Preston CJ, Sheinin MY, Sproat DJ, Swarup VP (2010) The novelty of nano and the regulatory challenge of newness. NanoEthics 4:13–26

Qi L (2008) Nanotechnology and nano-safety research must be synchronized. Science Times, 29 January

ReachCentrum (2008–2011) REACH e-learning package. http://www.learningspan.com/ReachCentrum/Safety_RC.html. Accessed 30 Nov 2014

Renn O, Roco M (2006) White paper on nanotechnology risk governance. International Risk Governance Council (IRGC). http://www.irgc.org/IMG/pdf/IRGC_white_paper_2_PDF_final_version-2.pdf. Accessed 30 Nov 2014

Reynolds GH (2003) Nanotechnology and regulatory policy: three futures. Harv J Law Technol 17:179–209

Rickerby D, Carbone AL (2008) Nanosystems for water quality monitoring and purification. www.oecd.org/dataoecd/36/29/42326650.pdf. Accessed 30 Nov 2014

Satariano A (2010) Pollution particles lead to higher heart attack risk. http://www.bloomberg.com/apps/news?pid=washingtonstory&sid=aBt.yLff. Accessed 30 Nov 2014

Shelley-Egan C (2010) The ambivalence of promising technology. NanoEthics 4:183–189

Shvedova AA, Kisin ER, Porter D, Schulte P, Kagan VE, Fadeel B, Castranova V (2009) Mechanisms of pulmonary toxicity and medical applications of carbon nanotubes: two faces of Janus? Pharmacol Ther 121(2):192–204

Sigma-Aldrich (2014) Globally harmonized system (GHS). http://www.sigmaaldrich.com/safety-center/globally-harmonized.html. Accessed 30 Nov 2014

Som C, Berges M, Chaudhry Q, Dusinska M, Fernandes T, Olsen SI, Nowack B (2010) The importance of life cycle concepts for the development of safe nanoproducts. Toxicology 269(2/3):160–169

Song Y, Li X, Du X (2009) Exposure to nanoparticles is related to pleural effusion, pulmonary fibrosis and granuloma. European Respiratory Society. http://www.ersj.org.uk/content/34/3/559.full. Accessed 30 Nov 2010

Staman J (2004) Foresighting the new technology Wave Expert Group SIG 2 final report V3.7 (11.7). ftp://ftp.cordis.europa.eu/pub/foresight/docs/ntw_sig2_en.pdf. Accessed 30 Nov 2014

Strand R, Kjølberg K (2011) Regulating nanoparticles: the problem of uncertainty. Eur J Law Technol 2(3). http://ejlt.org//article/view/88/157. Accessed 30 Nov 2014

Suttmeier RP, Cao C, Simon DF (2006) 'Knowledge innovation' and the Chinese Academy of Sciences. Science 312:58–59

Tang L, Carley S, Porter AL (2011) Charting nano environmental, health, & safety research trajectories: is China convergent with the United States? J Sci Policy Gov 1(1):1–16. http://works.bepress.com/li_tang/9. Accessed 30 Nov 2014

Tangwa G (2004) Between universalism and relativism: a conceptual exploration of problems in formulating and applying international biomedical ethical guidelines. J Med Ethics 30(1):65–8

The Center for International Environmental Law (2012) Nanomaterials "Just Out of REACH of European Regulations". http://www.ciel.org/Chem/JustOutofREACH_Feb2012.html. Accessed 14 Nov 2014

The International Risk Governance Council (2014) About IRGC. http://www.irgc.org/about/. Accessed 30 Nov 2014

The Royal Society & Royal Academy of Engineering (2004) Nanoscience and nanotechnologies: opportunities and uncertainties. http://www.nanotec.org.uk/report/chapter10.pdf. Accessed 30 Nov 2014

Tomellini R, Giordani J (2008) Third international dialogue on responsible research and development of nanotechnology. http://ec.europa.eu/research/industrial_technologies/pdf/policy/report-third-international-dialogue-2008_en.pdf. Accessed 30 Nov 2014

Trager R (2008) EPA nanosafety scheme fails to draw industry. Chemistry World. www.rsc.orgg/chemistryworld/News/2008/August/0508081.asp. Accessed 30 Nov 2014

Tsuchiya T, Oguri I, Yamakoshi YN, Miyata N (1996) Novel harmful effects of [C60] fullerene on mouse embryos in vitro and in vivo. FEBS Lett 393:139–145

UNESCO (2006a) The ethics and politics of nanotechnology. http://unesdoc.unesco.org/images/0014/001459/145951e.pdf. Accessed 30 Nov 2014

UNESCO (2006b) COMEST extraordinary session. http://unesdoc.unesco.org/images/0015/001514/151443e.pdf. Accessed 30 Nov 2014

van Caster G (2007) The role of the World Trade Organization in nanotechnology regulation. In: Hodge GA, Bowman DM, Ludlow K (eds) New global frontiers in regulation. The age of nanotechnology. Monash studies in global movements. Edwards Elgar Publishing, Cheltenham, pp 287–319

Visser R (2007) A sustainable development for nanotechnologies: an OECD perspective. In: Hodge G, Bowman D, Ludlow K (eds) New global frontiers in regulation. The age of nanotechnology. Edward Elgar Publishing, Cheltenham, pp 320–332

von Schomberg R (2010) Introduction. In: von Schomberg R, Davies S (eds) Understanding public debate on nanotechnologies: options for framing public policy a report from the European Commission Services. Publications Office of the European Union, Luxembourg. http://ec.europa.eu/research/science-society/document_library/pdf_06/understanding-public-debate-on-nanotechnologies_en.pdf. Accessed 30 Nov 2014

Widmer M, Knebel S (2009) Fishing for nano risks with the wrong type of fishing net: European Parliament calls for sweeping review of current regulations concerning nanomaterials. http://www.innovationsgesellschaft.ch/en/index.php?section=media1&path=%2Fmedia%2Farchive1%2F2009%2F. Accessed 30 Nov 2014

Wu J, Liu W, Xue C, Zhou S, Lan F, Bi L, Xu H, Yang X (2009) Toxicity and penetration of TiO2 nanoparticles in hairless mice and porcine skin after subchronic dermal exposure. Toxicol Lett 191:1–8

Zhao F, Zhao Y, Wang C (2008) Activities related to health, environmental and societal aspects of nanotechnology in China. J Clean Prod 16:1000–1002

Chapter 5
pTA (Participatory Technology Assessment), Habermas's Dialogue/Discourse Ethics and Nanofora

> *...to yoke the S&T behemoth to ends chosen by the people.*
> *(Bostrom 2007)*
>
> *At what stages in scientific research is it realistic to raise issues of public accountability and social concern? How and on whose terms should such issues be debated? Are dominant frameworks of risks, ethics and regulation adequate? Can citizens exercise any meaningful influence over the pace, direction and interactions between technological and social change? How can engagement be reconciled with the need to maintain the independence of science, and the economic dynamism of its applications? (Kearnes et al. 2006)*

Abstract This chapter looks at the pTA 'nanofora' that have been attempted thus far in the EU and China (the latter being somewhat less concerned with this approach as yet) and links pTA to the field of discourse ethics. The principles Habermas sees as requisite for dialogue are very similar to those of pTA, particularly those of inclusivity and range of representation. Habermas asks the basic question of global ethics, of how different views (particularly of social order) can be universally understood and reconciled, perhaps within an 'ideal community' of communication, one that may be global. His 'universalisation principle' offers a useful methodology for examining how a pTA model might work globally.

Keywords pTA • Habermas • Discourse ethics • Universalisation principle

The first half of this study has offered a comparative analysis of the social, bioethical as well as policy contexts in which nanoissues are identified and regulated in the EU and China. The conclusion to this first section suggested that risk issues are being addressed in similar ways by both regions, although in the EU the precautionary principle may offer greater stringency, and there is a situation of 'hybrid' regulation in that region. This means a mix of voluntary reporting/codes of conduct (criticised for ineffectiveness) and identification and regulation of nanomaterials under existing regulations (criticised for lack of regulatory reach and 'loopholes').

By contrast, China has a top-down or closed regulatory system.

Global bodies, as shown in the previous chapter, are most effective at clarifying standards for nanomaterials through existing chemical regulations. None have achieved much forward momentum though, either in terms of the broader social or ethical issues relating to nanotechnology.

In short, the societal ethics of nanotechnology remain unaddressed in policy terms. This would seem to be a reason to involve society more closely in policymaking.

This brings us to the question, not directly addressed thus far, of *who exactly* should be involved in global regulation and policy.

On the assumption that global ethics are enacted by global agents, the following three chapters look at *how* global agents might achieve consensus, i.e. through which virtues or skills, and by means of which procedures. We will look at the logic behind this assumption by asking the following questions:

- Why agency?
- Why might *lay* agency, through pTA, be a preferred model for building S&T dialogue?
- What does pTA actually involve?
- And why do ideas from Jürgen Habermas's discourse ethics offer a usefully extended model of pTA as a tool for dialogue?

5.1 Agency as a Global Idea?

Globalization has opened new opportunities for participatory politics. (Delanty 2000)

Of course, the phrase 'global idea' – implying some form of universal consensus on values – is immediately contentious. *Is* there such a thing as a universal value?

While we may not agree that 'consciousness of temporality is the fundamental, defining attribute of human beings…respect for the time of the other should be the modality in which responsibility for the other can be generalized in the public spheres' (Venn 2002), we may wish to agree that there are indeed values, directives or aims that should be promoted and protected on a global basis. But which ones? In his 1990 book *Global Responsibility: In Search of a New World Ethic*, Küng argues for a globally shared ethics based on religious directives such as non-violence, respect for life, solidarity, just economic order, tolerance, truthfulness and equality.[1] Tremblay (2010) suggests ten 'principles for global humanism' – dignity, respect, tolerance, sharing, no domination, no superstition, conservation, no war,

---

[1] See http://www.weltethos.org/dat-english/01-history.htm. Küng arguably and ironically implies one of the particular issues working against a global ethic; local, historical and cultural determinants such as religion, allied to an often extremely emotive view of 'truth'. However, Kimberley Hutchings (2010) looks at Küng's claim 'that we already have the resources to address the *why, what, who* and *how* questions of Global Ethics within existing world religions'.

democracy and education. And Fox (2006) argues for a foundational ethical principle – that of the 'responsive cohesion' of values that through their mutually modifying interactions generate an overall cohesive order – while Burmeister et al. (2011) suggest 'equality, freedom, respect and trust'.

It seems that there might be some similarity in these universal aims, yet no certainty on any *hierarchy* of them.

To try and find an ideal or value on which there would be agreement is not easy – even if we take the very basic right, that of the right to life, as an example. Most would surely agree that murder is bad, but do they all feel the same way if they are supporters of capital punishment, or anti-abortionists, or those waging *jihad*?

The argument between pragmatists and those who, like Amartya Sen (1980), believe that systems fail through ignoring 'non-utility information' (i.e. religious or other ideological matters instead of basic needs such as food and shelter) is not new.

This discussion could be extended infinitely, but the premise here is that there is unlikely to be universal agreement (certainly, there isn't much current evidence in the world as we know it). Therefore, the first point to be made (unsurprisingly, given the emphasis placed later on Habermas's ideas as centred on procedure) is that any 'new universalism' is more of a question about agreed *method*, rather than agreed goals, principles or values.

Thus, Ladikas and Schroeder (2005) suggest that global ethics 'is not a field of academic study, it is an activity; the attempt to agree on fundamental conditions for human flourishing and to actively secure them for all', and is reliant on such practical activities as platforms for intercultural dialogue and trust-building, as well as international ethics committees and ethics reviews for ongoing global negotiations. Any solutions to the 'issue' of global ethics, surely, should include an implementation element. There is a need for a procedural approach. Pogge (1999) argues that as: 'disagreements about what human flourishing consists in may prove ineradicable, it may well be possible to bypass them by agreeing that nutrition, clothing, shelter, and certain basic freedoms, as well as social interaction, education, and participation, are important means to it – means which just social institutions must secure for all'.

Pogge's view implies basic citizenship rights (survival provisions), a general democratic state of choice (to improve oneself, e.g. through the freedom to receive education) as well as societal interaction. However, these rights are not globally accepted, as in the right of women in certain countries to choose education. Pogge's formulation also turns out to be fairly broad, but in its mixture of practical suggestions (shelter) and more idealistic ideas (freedoms including that of participation), he outlines one of the main issues with global ethics: that very mixture of the pragmatic ideas with 'value-added' ideas on which there is less agreement. For one might reasonably argue that all humans have the right not to live on the street (although strangely, they still seem to in most countries), but not necessarily agree that everyone should have equal democratic participation.

This also leads us beyond a strictly procedural emphasis, to one focused on the agent; ethics is implemented by institutions and by agents within certain procedures.

There is an obvious problem of implementing moral systems – namely, that they need to take 'account of messy realities on the ground in charting a practical course towards that objective' (Arras 2010).[2] Universal values are replaced by agents' desires to achieve compromise. Thus, Seyla Benhabib (1993) argues that discourse ethics is Kantian ethics 'collectivised'. Instead of asking what an individual moral agent could or would will, without contradiction, to be a universal maxim for all, one might ask:

> what norms or institutions would the members of an ideal or real communication community agree to as representing their common interests after engaging in a special kind of argumentation or conversation? The procedural model of an argumentative praxis replaces the silent thought-experiment enjoined by Kantian universalisability.

Rawls (1971) saw advantage as based on means (income, primary goods) and on 'goods' that included basic rights and liberties, freedom of movement and of choice of occupation, income and the social base of self-respect – a platform of means indispensible for realising certain democratic ideals. Sen gave this argument a different emphasis, seeing the significant issue as what 'goods' enable people *to do*. Sen's (1999) capability approach is defined by its choice to focus upon the moral significance of an individual's capability of achieving the kind of life they have reason to value.

Agents are not abstract, as Spence (2007) defines them, but are any socially engaged actors who 'value their freedom and wellbeing precisely because they recognize them as being the necessary enabling conditions for the fulfilment of their own specific individual and communal purposive actions'. The above definition is useful for two reasons. First, it picks up on the notion of freedom, linked to wellbeing, but also to the contentious issue for universalism, that of individual human rights or freedoms. Second, it relates to the notion of *dual* agency – one's own actions and 'communal' actions (i.e. actions undertaken to achieve common rather than individual aims).

By freedom, one might mean Dissanyake's (1996) 'I conceive of the human agent as the locus form in which reconfirmations or resistances to the ideological are produced or played out' or, more usefully, freedom as the ability to achieve. Sen's (1999) capability approach, in which the focus is on the actual freedom a person can exercise to choose 'doings and beings', argues that agency freedom must be present for the agent to advance in well-being; thus, greater freedom of agency implies a larger capability set. In 'What are capabilities?', Martha Nussbaum (2000) suggests ten capabilities, arguing that justice demands the pursuit, for all citizens, of a minimum threshold of these ten capabilities. These are:

- Lifespan
- Bodily health (and nourishment)

[2] Even Rawls distinguished between ideal and non-ideal situations, although he did not fully address the question of by what moral standards a person should govern his/her conduct given a reasonable level of compliance.

- Bodily integrity (including freedom of travel, sexual security and choice in matters of reproduction)
- Senses, imagination and thought (including the right to education and freedom of expression)
- Emotions (being able to have attachments to things and persons outside ourselves, along with not having one's emotional development blighted by fear or anxiety)
- Practical reason (being able to form a conception of the good and to engage in critical reflection about the planning of one's own life)
- Affiliation (social interaction, equality, dignity)
- Other species (being able to live with concern for and in relation to animals, plants and the world of nature)
- Play (being able to laugh, to play, to enjoy recreational activities)
- Control over one's environment (both in terms of political choice and material goods) (Nussbaum 2000)

Thus, while Sen, supposedly, 'filled in the moral space of functioning and capability', Nussbaum (1988) 'fills in the picture by identifying those "central functional capabilities" that are (allegedly) necessary and sufficient for the good human life'.

Two interesting aspects of Sen's theories are that they imply democracy and that they allow for cultural relativism. In terms of democracy, Korsgaard's argument suggests displacing the responsibility for agency onto the agent.

Rather than contentious phrases such as the '*good* life', given that there are many concepts of such a life, we might note Nussbaum's addition of reason and affiliation in her list of capabilities, which allow the capable person entry into a *moral* life as Habermas might define it; it is a life of rational, social reasoning towards consensus.

The phrase 'good human life' indicates an issue with the capability approach; agency needs to be directed. As Crocker (2008) argues, agency requires teleology in that they must have a purpose, which Crocker then sees as a concern for others. This surely implies some 'universal value' towards which agents are directed. (This is not always logical, given that one's teleology could simply be a concern for one's self.)

The teleology of the agent in terms of capability is his/her ability to develop the rational conditions for his/her own flourishing – and for that of other people. Korsgaard et al. (1996) has a version of this argument:

> Morality is grounded in human nature. Obligations and values are projections of our own sentiments and dispositions. To say that these sentiments and dispositions are justified is not to say that they track the truth, but rather to say they are good. We are better for having them, for they perfect our social nature, and so promote our self-interest and flourishing.

Sen (1993) suggests a dual teleology. His model allows for a distinction to be made between personal and 'altruistic' goals, arguing that:

> …when more capability includes more power in ways that can influence other people's lives, a person may have good reason to use the enhanced capability – the larger agency freedom – to uplift the lives of others, especially if they are relatively worse off, rather than concentrating on their own well-being.

Sen's view is that one 'may have good reason' to act altruistically, if one has the capacity to do so. Sen (1977) asserts that people can make choices that are not in one's self-interest, referring to this as a 'commitment', where an individual chooses an action, say, for reasons of duty, even if it affects personal self-interest. (To offer another example of this belief, in his environmental ethics, Sandler (2007) notes 'ends independent of our own flourishing' that may also seem 'good' and rational, duties that may in fact 'trump' self-interest.

It can be argued that Sen implies selfish altruism – acting out of duty enhances one's self-image and probably one's standing in the community – but this is of less relevance than the notion that one might operate on differing levels of capability, with one focused on self and the other on a more public self.

Sen makes another useful point by allowing for adaptation to context and to culture in terms of determining the capability set of a person, i.e. that agency allows for a recognition of cultural pluralism or relativism. Cultural agency is another capability.

Agency freedom is seen both as an individual and a communal value. This is not too far from the ideas of Gewirth (1984), who sees any agent as able to act in a universal manner. Gewirth's ethical rationalism offers a 'principle of generic consistency' in which every agent must act in accordance with his or her own generic rights to freedom and well-being in addition to that of all other agents – the maxim being, 'act in accord with the generic rights of your recipients as well as of yourself'.

Gewirth's normative force apparent in any action is procedural, like Habermas' impulse towards consensus, thus escaping the issue of universal values since Gewirth grounds ideas of human rights in generic features of action such as voluntariness and purposiveness common to all agents. For Gewirth (1982), action or agency supplies the metaphysical and moral basis of human dignity and personhood, providing human rights with a kind of 'self-grounding' – 'the latter might also be termed "natural rights" in that they pertain to humans simply in their capacity as actors or agents'. (It can also be noted that dignity, in Gewirth's theories, derives from agency, the preferred term for discussing individual human rights and individualism in Chap. 2, in that it suggests a universal value as well as an individual one (White 1982).)

Agency is defined in this study firstly as procedural, as requiring a structure for action, and so we will examine nanofora in this context. Second, it is an aspect of one's public or private identity. This is not too far a stretch from the notion that one may have both a private cultural self and a public, transcendental or universal self, an idea to be developed in Chap. 6.

Before looking at nanofora, i.e. the structures that might allow the agent to exercise agency, hopefully globally, there is one further aspect of agency to be considered – that of *lay* agency. Why might we focus on lay agents rather than scientists or policymakers? In other words, why is the preferred agency model that of pTA?

5.2   Technology Assessment (TA) and Participatory Technology Assessment (pTA)

> Technology assessment…aims at clarifying socioeconomic and environmental problems potentially attendant on technological developments and thereby providing information to the public and the government … that will inform public-policy decision making as well as guide R&D planning and natural-resource allocation. (O'Brien and Marchand 1982)

pTA can be defined as a consultative process through which the public's opinions are solicited. It involves various kinds of social actors such as different kinds of civil society organisations, representatives of the state systems, individual stakeholders and citizens (laypersons). (The narrower definition of pTA is purely 'layperson's opinions'.)

The increasing interest in pTA reflects a shift from

> a largely closed, intrainstitutional tool of policy analysis and advice to a tool for the social assessment of scientific-technological issues at the interface between politics and public discourse. Through citizens' conferences, scenario workshops, and consensus conferences, technology assessment has effectively been opened up to the public sphere: Citizens and interest group representatives are drawn into the process of assessing scientific and techno-logical issues alongside experts, the process often takes place in public, and its outcomes are made widely available for information and debate. (Joss 2002)

pTA, which I hypothesize as a version of the Habermasian model of discourse ethics, suggests a significant role for the public in global nano(ethics). Is this justified? Who is involved in technology assessment and policymaking, and how do these spheres of agency overlap? Governmental regulators, when formulating public policy on S&T, draw on a range of expert advisors from the scientific and other communities. The following offers a brief overview of how technology assessment has developed since its rise in the 1980s.

Where in the cycle of research and product manufacture does TA begin? The ethics of researchers, corporations and those bodies funding research requires questions of consequence and methodology to be addressed during the developmental stage. For example, stakeholder analysis, focusing on the notion of 'mediation', or analysis of the future role played by technology in human actions and experiences suggests the simple approach of bringing societal concerns to the fore at the earliest possible stage in technology development. Thus, it requires the effective involvement of as broad a range of members of society as possible. The 'value-sensitive design' approach, first proposed in the context of information and communication technology, suggests a dual approach by the engineers of products and its external stakeholders or users. 'Value-sensitive design' aims to connect the people who design products and systems to the stakeholders affected by these products.

In the value-sensitive design approach, the core notion is, simply put, one of dialogue between users and developers leading to a translation of societal concerns into products. Yet the situation is arguably more complex, as Jotterand (2006) argues when he states that:

> …analysis of the social and ethical implications of nanotechnology must confront the pluralism of scientific discourses and the plurality of values and norms these discourses entail

... (post-academic science) ... is characterized by complex relationships between the industry, science, academia, economics, and politics that determine what type of research and development ought to be pursued on the ground of ethical, economic, political, and social reasons. (This) requires a dialectic between diverse fields within science and technology as well as between the sciences/technologies and the humanities.

In other words, the broader and more pluralistic modes of assessment can be the best ones. This means that various types of TA, pTA being merely one of them, should be encouraged. 'Constructive TA' offers user feedback to researchers, while 'expert TA' is usually confined to scientists but can be combined with pTA in mixed expert/layperson fora (while parliamentary TA operates through the medium of parliamentary advisory fora).

Why pTA? Before answering this question, it should be noted that there is also a 'why not', in that pTA is often seen as ineffective. The cynical view is that it is merely a cosmetic exercise. pTA offers a way for governments, corporations and research institutions to assess community needs and reactions, but that does not necessarily translate into substantive changes in policy, but only to a sense of public 'oversight' in the broadest sense. Thus, pTA only has a 'limited effect on political decision-making' (Abels 2007).

Joss (1999) sees pTA as increasingly recognised by public institutions – though recognition is not the same as implementation. This statement might be justified by examples such as that of the 2009 'Nanopodium' project run in the Netherlands, a public dialogue about the threats, opportunities and applications of nanotechnology. Although its report was submitted to the Dutch parliament in 2011, it remains to be seen what effect this might have on policy. (In 2009, the Lower House of the Dutch parliament introduced three motions concerning nanotechnology; these however were limited to risk analysis issues rather than any societal or ethical concerns.) One outcome has been the suggestion to increase NGO involvement in policymaking (van Est 2008).

In the UK, three nanodialogue projects (NanoJury UK, Small Talk and Nanodialogues) were intended to inform nanotechnology policy, and all reported their findings to the government and other relevant institutions. Yet, according to Blum (2012), only one institution has responded formally to these projects.

The question is how to link the public's voice with the policymaking process. Dryzek and Tucker (2008) suggest that there is potential for linking pTA to established electoral processes. Having

a permanent channel of consultation and communication between parliamentarians and the public would significantly improve public trust in the governance of science and technology .... A national institution could, in the right circumstances, and with enough political support, guarantee that public dialogue has policy impact. (Sciencewise 2011)

The adoption of the EU 'Common Position' on armaments[3] was allegedly a demonstration of the fact that 'a well-organized citizen's lobby can achieve a great deal'

---

[3] For a discussion of what this entails, see http://www.sipri.org/research/armaments/transfers/controlling/eu_common_position.

(McGiffen 2005). The European Citizens' Consultations offer other examples of EU-wide deliberative democracy. (And 2013 is the 'European Year of Citizens', aiming 'to mobilise and coordinate wide civil society engagement' and 'initiate a European-wide debate on issues relating to the exercise of European citizens' rights and to citizens' participation in the democratic life of the EU' (European Year of Citizens 2013 Alliance).)

There is also some concern about how democratic pTA truly is. As Lewidow (2009) outlines, there is a need to extend the agenda for, and ownership of, such fora, for 'the prospects for democratization depend upon wider, autonomous forms of participation – neither sponsored nor welcomed by state bodies'. In other words, pTA runs the risk of educating the public sufficiently to understand that they might be being cajoled into an exercise that offers the comforting illusion of democratic process.

However, the negative view of pTA is a limited one. The 'effectiveness' of pTA can be measured in terms of more indirect and informal influence. pTA is useful for global ethics and global policymaking in the context of new technologies for the following general reasons:

- It offers new, broader and pluralistic methods of governance for the twenty-first century and for a twenty-first-century technology like nano, reflecting the growing sense that technological innovations are seen as the outcomes of social networks that incorporate a range of actors (Nowotny et al. 2001).
- It addresses the public trust deficit associated with many new technologies and such a deficit's attendant economic and political problems.
- It provides a method of awareness raising, particularly with complex new technologies that may not be communicated effectively by governments or the media, thus leading to better informed citizens.
- It is an effective method of promoting self-reflection and bridge-building among actors with conflicting views in S&T debates.

There are more arguments that support pTA when seen as a development of classical expert-focused TA (not least in relation to the lack of understanding of socio-ethical issues within the expert community themselves), but as the focus in this study is on what pTA stands for rather than the specific merits of one methodology over another, that issue can be left for future empirical studies.

Given the hypothesis that will be tested over the following chapters, that global ethics requires individual agency at both national and international levels or a 'dual identity' as I shall term it, policymaking needs to be broadened from national committees to global citizens. The skill set required of a global citizen will be outlined in the following two chapters, but the catalyst for such skill development is public empowerment. In other words, a more deliberative, democratic form of engagement with S&T ethics is required. Global ethics necessitates global agency, which requires global empowerment. The latter, given differing political structures, is predicated as individual or lay empowerment through pTA.

5.2.1   The Public Trust Issue

The complexity of public decisions seems to require highly specialized and esoteric knowledge, and those who control this knowledge have considerable power. Yet democratic ideology suggests that people must be able to influence policy decisions that affect their lives.
(Nelkin 1975)

The power of the consumer is ever increasing. Even in less democratically inclined
nations. governmental policymakers and regulators are increasingly aware that
there is a strong economic argument for greater public involvement in 'approving'
new products for market; the trust deficit is bad for business, as was brought home
with the billion-dollar losses incurred through the GM boycott. Scandals,
particularly those relating to food, have left a legacy of suspicion that has fed into
the EU nanodebate, giving rise to a fear that nanotechnology might be 'the next
GM' (Walsh 2009; Kyle and Dodds 2009).

Public dissatisfaction in China after various food safety scandals relating to milk
and to gutter oil suggests scepticism about regulation is not only an EU issue. The
2008 scandal, when milk from the Sanlu Corporation was found to be contaminated
by melamine, was China's second major baby-milk scandal. Reports of the death
toll from the Sanlu contamination vary according to whom one reads, from 3 to 11,
but up to 300,000 children were reported as affected to some degree. The gutter oil
scandal erupted in September 2011 after reports about companies recycling oil from
drains behind restaurants first appeared. In China, policymakers are beginning to
become more concerned, adopting a more inclusive approach, 'a more reflective
approach to engaging with the public', that is, 'real dialogue' (Jones 2009).

Boycotts of some products have not only had an economic impact for corporations; public attacks on governments undermine governmental authority (Kearnes
et al. 2006).

There is a growing view that policymaking should not be undertaken purely by
government regulators, due to growing public scepticism about the lack of neutral
experts advising on S&T policy, thus creating a 'de-legitimisation' of policy. As
Porter et al. (1980) argue,

A technology cannot be described or forecast without reference to society. Society contributes significant public resources to technologies seen as leading to desirable social goals. It
has also demonstrated a willingness to withhold support from technologies it deems
undesirable.

5.2.2   The 'Legitimisation' Effect of Public Participation

Scandals where products have affected human health are now more quickly and
widely disseminated, due to increased global media, with an accompanying outcry
against regulators and governments. This has led to a demand, both for more
informed public debate and for greater public involvement in how new technologies

should be regulated and funded, as well as how they should be applied to everyday life (Gaskell et al. 2002). (A note of caution is sounded by Schummer (2004), who, referring to the USA's 2003 National Nanotechnology Initiative, which prioritises S&T that 'brings about improvements in quality of life for all Americans', argues that societal concerns should balance, not *impede*, nanodevelopment.)

The challenge consists in closing the widening democratic gap between policy-makers and citizens, thereby increasing the citizen's sense of ownership in decisions made about current global issues. Policymakers now call for the inclusion of discussion of the ethical and societal impacts of potentially transformative new technologies at an early stage in policymaking – the so-called upstream engagement (Friedman 1997). The term 'upstream engagement', which 'views the trajectory of new technologies as being decided not solely by the scientists and industrialists who seek to develop it, but rather with governments having a responsibility to engage the public in the decision and policy-making processes' (Hodge and Bowman 2007), is another way of defining pTA (Siegrist et al. 2007). Thus, 'setting up public panels has already become an accepted method within ELSI-research programs' (van Est 2011). As Horst (2010) argues,

> Developments for example in biotechnology have sparked off a number of social controversies during the past decades and it has been common to understand public debate as a necessary prerequisite for the ability to deal with these controversies. (There is) broad discursive consensus in favour of public debate and participatory exercises regarding the social responses to biotechnology.

As Kaiser et al. (2009) discusses, with public participation, societal expectations can be unmasked and exposed through research (interviews); such 'identified expectations are fed back into the public dialogue as contestable values or disputable attitudes that need to answer to accountability requirements'. By subjecting such expectations to scrutiny and insisting on accountability, the risk/benefit debate can be continuously adjusted. This all may mean a re-evaluation of the public's ability to tolerate risk (and so to accept new nanoproducts); the 2006 Swiss canvassing of public opinion on nanotechnology in food products and packaging discovered that lay people considered risks to be of more importance than scientists did. In addition, upstream engagement presents the:

> …potential for engaging with new questions beyond those traditionally employed in discussion about risks … ideally, this dialogue would explore not only visions for society…but assumptions about the ways these visions are constructed. A key conclusion here is that 'upstream' public engagement…must move beyond conventional 'risk communication' based dialogue, to be future focused, broadly framed, and to explicitly incorporate questions of both public values and technology governance…all of this raises significant challenges. (Pidgeon and Hayden 2007)

This is all particularly significant for nanotechnology. One of the 2006 EU-funded DEEPEN (Deepening Ethical Engagement and Participation in Emerging Nanotechnologies) project aims was that of 'developing methodological tools for engaging civil society and the nanoscience community in ethical reflection', hoping to 'integrate understanding of the ethical dilemmas posed by emerging nanotechnologies into the innovation trajectories of the technology itself' (European

Commission 2010). We have a 'rare opportunity to integrate societal studies and dialogues from the very beginning' in the nanodebate (Rocco and Bainbridge 2001). Nanotechnology, the 'natural inheritor of a complex web of disenchantment, tension and ill-feeling caused by a series of recent technological controversies in Europe' offers 'an opportunity to road-test new forms of participatory and deliberative public engagement with a technology in the early stages of its development' (Macnaghten et al. 2005).

Jürgen Habermas (1991), discussing the 'legitimation crisis' that governments may suffer from when public trust is lost, argues for greater power in the public sphere, something he describes as:

> a warning system with sensors that, though unspecialized, are sensitive throughout society. From the perspective of democratic theory, the public sphere must, in addition, amplify the pressure of problems, that is, not only thematize them, furnish them with possible solutions, and dramatize them in such a way that they are taken up and dealt with by parliamentary complexes.

In his *Between Facts and Norms* (1992), Habermas presented a social theory that addresses the tension between moral norms and their practical context through legal institutionalization based on discursive procedures and the principle of democracy – thus 'only those statutes may claim legitimacy that can meet with the assent *(Zustimmung)* of all citizens in a discursive process of legislation that in turn has been legally constituted'.

Collins and Evans (2007) describe the legitimacy problem succinctly, stating that at the start of the twenty-first century 'it is well established that the public have the political right to contribute, and without their contribution technological developments will be distrusted and perhaps resisted'.

There is arguably a greater sense of 'de-legitimisation' that derives not merely from governments, but from the condition of modernity itself. Hennen (1999), for example, argues that pTA can be analysed in the light of current sociological debate about 'uncertainty' and that pTA, 'as a response to technological controversy, should be understood as a means of dealing, in creative and interactive ways, with the issue of (scientific, social, ethical…) uncertainty at the heart of modern society'. This indicates that 'knowledge' in an age of certainty has become less firmly owned by any one body or institution; however, the means for achieving an ethical discussion of 'technological controversy' may require a new and 'decentralised' form of approach.

The extension of S&T debate into the public sphere, which Habermas (1996) describes as an autonomous space where citizens engage in reasoned discourse, helps to strengthen the foundations of democracy; more specifically, it plays a role in strengthening 'democratic will-formation'. Habermas acknowledges that deliberative democracy needs to be institutionalised (must become part of policymaking), yet he also sees it as a significant part of the ongoing process of decentralisation. It encourages citizens to work towards the kind of society they would like to live in. In this, it extends the idea of pTA or discourse events as learning processes not only in terms of the communication of data and ideas but also as a means of developing democratic skills.

The issue of decentralisation, as I shall argue in Chapter Six, is a crucial requirement for global ethics, in that it allows discourse to enter into a new public space.

5.2.3   New Forms of Governance?

Science issues, in their complexity, and in the trust deficit context, have become 'a major challenge for governance'. (Fuller 2000)

As Abels (2010) explains, the notion of governance doesn't solely mean institutions anymore, rather 'formal and informal aspects of co-operation and coordination between a diversity of social actors… new modes of policy-making by linking the input (participation) and output (effectiveness) dimensions of democratic legitimacy'. There is a growing sense that technological innovations are seen as the outcomes of social networks that incorporate a range of actors (Nowotny et al. 2001). The social and ethical issues that S&T raise are no longer the responsibility of a 'relatively closed bureaucratic-professional-legal world of regulation', but are relocated within the public arena (Jotterand 2006).

It is already expected that policymakers and other governmental regulators will consult with think tanks, NGOs, researchers and other S&T advisory bodies; the layperson's input has become increasingly important in TA and indeed was part of the initial TA landscape in some European countries. pTA was also part of the move towards greater democracy and the trialling of consensus conferences for greater public discourse, particularly in Europe during the 1980s (Brom and van Est 2011).

The Danish Board of Technology (*TeknologirÂdet,* or DBT), for example, has a responsibility to encourage public debate and to serve as an independent source of advice and assessment on technology issues for the Danish parliament. In the DBT's mission statement, special emphasis is placed on clarifying the interaction between technology, society and people; the DBT strives 'to ensure that technology in Denmark is in harmony with the desire for a democratic, fair and economically, ecologically and socially sustainable society' that makes use of 'expert knowledge as well as the insight, experience and credibility of non-expert citizens'.[4] While KIT (the Karlsruher Institut für Technologie) in Germany is organising the Citizen's Dialogue on S&T, the Norwegians, Austrians and Swiss have similar projects, and many other TA institutes are now focusing on lay participation methodologies. The Rathenau Instituut in the Netherlands is known for the methodology of horizon scanning, or foresight, a process in which laypersons are often included. In their September 2012 report, the institute argued that nanotechnology and pTA should operate in tandem, for:

Nanotechnology has provided a new window of opportunity to reframe state-science-society relationships. In particular the notion of upstream public engagement has been put forward…Our study shows that in order to better understand the complexities of the governance of science and technology, a new research perspective is needed [for] reflecting on

---

[4] See http://www.tekno.dk/subpage.php3?page=statisk/uk_about_us.php3&language.

the relationship between informing and engaging, on the interaction between engagement processes within the societal, scientific and political sphere, and on organisational and institutional constraints. (van Est et al. 2012)

The UK's Foresight Horizon Scanning Centre, similarly, offers as part of its mission the statement that 'at a time of substantial public service reform, futures thinking has a critical role to play in helping to ensure that wider perspectives are identified, traditional assumptions challenged and new possibilities explored'.[5]

5.2.4 *Educating Citizens and Policymakers*

The pTA process educates policymakers and the public. Sclove's 2010 Wilson report on the need for a 'reinvented TA' in the USA argued that 'informed groups of citizens could identify a wide and rich range of issues associated with new technologies, often adding nuances to the views of experts and the policymaking community' (p. vi). Sclove proposed a new national expert-and-participatory TA institutional network – the Expert & Citizen Assessment of Science & Technology (ECAST) network. ECAST would be independent of the government and comprise a complementary set of non-partisan policy research institutions, universities and science museums across the USA.

Broader procedures incorporating more actors might increase the public knowledge base and produce 'new possibilities of conflict resolution' and 'realise common interests' (Bora and Hausendorf 2010). Such twenty-first-century methodological pluralism solves the problems that Ladikas (2009) notes in the context of Science in Society (SIS) programs – for example, issues of political will, the narrow framing of science issues, methodological pluralism and knowledge integration – all problems that pTA aims to resolve.

This pluralism also places pTA at the forefront of global engagement with S&T.

Public awareness of, and familiarity with, nanotechnology is still low, according to a 2004–2007 study across the USA, Switzerland, Japan and Europe (Satterfield et al. 2009). One of the ways in which the accusations of pTA's inefficacy can be countered is to redefine the notion of 'impact'. This can be widened to take into account a TA event's 'resonance', by which it is usually meant awareness raising and action initialising. Thus, Guston (2011) argues that pTA should be assessed qualitatively rather than in terms of effectiveness:

…pTA offers a combination of intensive and extensive qualities that are unique among modes of engagement … this combination led to significant learning and opinion changes, based on what can be characterized as a high-quality deliberation. The quality of the anticipatory knowledge required to address emerging technologies is always contested, but pTAs can be designed with outcomes in mind – especially when learning is understood as an outcome.

---

[5] See http://www.bis.gov.uk/foresight/our-work/horizon-scanning-centre.

Thus, Loeber et al. (2011) note that impact studies of pTA tend to focus on its effect on policy and media, without necessarily looking at the valuable contribution it makes to the arenas of 'public contestation'. In their view,

> the very staging of a participatory technology assessment implies the creation of a new space for contention in which relationships between actor-networks may shape up, evolve and change, or alternatively, reconfirm identities and boundaries.

In other words, pTA not only decentralises legitimation from the 'top' to the public but encourages a process of agency in which identity might shift its boundaries. This point will be discussed in more detail in Chap. 6.

The NEG report mentioned earlier notes that public engagement has a valuable impact in terms of contextualising science policy, research and governance, creating better informed, scientifically aware citizens while overcoming preconceptions (Gavelin et al. 2007). In addition to education, this report proposes that one might look at effectiveness differently, in terms of influencing researchers and decision-makers to consider the social context of their work more carefully:

> Direct links between public engagement activities and decision-making rarely happens ... instead we believe that public engagement activities are more likely to influence policy and decision-making through more subtle and indirect avenues ... a public engagement activity may challenge the views and attitudes of those who take part, thus leading to a gradual change in the priorities of decision-makers or researchers. (Gavelin et al. 2007)

5.2.5   Self-Reflection and Bridge-Building

Are pTA methods transferable across countries? Are there any current examples of East-West dialogue that provide a basic model? In 1996 the first Asia-Europe Meeting (ASEM) with European leaders and ten heads of state in Asia was held. While there is potential for a more political agenda within the ASEM framework, including human rights and security questions, it is unclear how those interests will be balanced with the economic priorities of both regions. The China-UK Economic and Financial Dialogue, founded in 2008, is the only vice-premier level economic and financial dialogue between China and Europe and was set up to develop long-term strategic relations between the two regions. Fang argues that 'the dialogue is still rudimentary and faces a lot of difficulties, but if these are overcome, it might produce a historic and world-class contribution to the global economy' (Fang 2011).

An overview of the dialogues between China and the EU suggests a 'complementarity of interests' in the two regions, so that:

> China and Europe have considerably more in common than might appear at first sight. This creates a strong mutual interest to promote the exchange of experience and know-how. China today is experiencing challenges which Europe started to tackle a number of years ago in areas such as the environment, the internal market, and competition. The EU is demonstrating its willingness to share this experience with China. And China has shown an interest in using the best practices of the 'EU model' in these policy areas. In other areas too, both Europe and China are simultaneously confronted with new challenges, such as

rapid advances in science and technology and problems with health protection. This is a two-way street. In some areas Europe could usefully benefit from Chinese know-how and experience. Peaceful nuclear research is an example of such an area, where Europe will soon have to close down its ageing experimental nuclear reactors, whereas China is currently building state-of-the-art facilities. (European Union External Action)

There are many such platforms for strategic discussion. What about S&T dialogue though?

EU-China S&T cooperation started in the early 1980s, through the European Commission's Framework Programme for Research and Technological Development, with a S&T Cooperation Agreement being signed in 1998. The 6th Framework Programme (2002–2006) supported 214 research projects involving Chinese teams. (FP7 (2007–2013) is furthering this successful cooperation.) The 2006 China-Europe Science and Technology Year was an example of a joint initiative of the Chinese Ministry of Science and Technology (MOST) and the Directorate-General of the European Commission. Horvat and Lundin's (2008) report on this cooperation also notes activities such as the exchange of equipment in addition to information, interviews and project pooling, with initiatives such as CO-REACH and BILAT-SILK.[6] These initiatives allow for Chinese partners 'to be more actively involved' and foster greater 'reciprocity and increased convergence of European and Chinese priorities' (Horvat and Lundin 2008). However, we may observe that such initiatives usually operate at the level of government or scientific community. There is less community involvement, although this should be essential, as argued by the Global Dialogue Foundation (GDF, established in July 2011 to promote intercultural understanding). The GDF predicates a two-level approach, one international and the other public, in pursuit of its aim to increase understanding and cooperation among people of different cultures, faiths and beliefs at the community-grassroots level, but also at the international, UN level. The project involves establishing networks of civil society organisations in each country.[7]

Examples of cross-national participatory TA projects are:

- Meeting of Minds (2004–2006)
- World Wide Views on Global Warming (2009)
- World Wide Views on Biodiversity (2012)

The Europe-wide pTA on advances in brain research, *Meeting of Minds*, a 2-year pilot project (2004–2006) led by a European panel of 126 citizens, was a concerted attempt by leading organisations in the field to move pTA and foresight to the European cross-national level. It gave European citizens

a unique opportunity to learn more about the impact of brain research on their daily lives and society as a whole, to discuss their questions and ideas with leading European researchers, experts and policy-makers, put them in touch with fellow citizens from other European countries and make a personal contribution to a report detailing what the people of Europe

---

[6] CO-REACH is dedicated to developing cooperation between the EY and China. BILAT-SILK is an EU Framework 7 programme that raises the EU's S&T profile in China.

[7] See http://www.globaldialoguefoundation.org/what_we_do.htm>l.

believe to be possible and desirable in the area of brain science and what they recommend policy-makers and researchers to be aware of for future developments in this field. (Slob et al. 2005)

The report on the project concluded that it demonstrated the feasibility, the effectiveness and the efficiency of public participation at a multinational European level.

In 2009, the Danish Board of Technology organised *World Wide Views on Global Warming*. Over 4,000 people in 38 countries met on a single day to discuss recommendations to an upcoming conference on climate change. Participants were selected for representativeness of age, education and other regional demographics. This event serves as an example 'of how to include everyday citizens in future global policymaking, giving them a bigger sense of political ownership and policy-makers a better insight in the views of the citizens they represent' (Silverman 2010).

And on Saturday, September 15, 2012, thousands of people around the world took part in *World Wide Views on Biodiversity,* a project that engaged ordinary citizens from 25 countries in policymaking to address the decline of biodiversity.

Given that a pTA approach seems useful for a new technology such as nano, based on the four reasons outlined above, how does one achieve public involvement in practical terms? The following takes a brief look at current pTA achievements in both the EU and China.

5.3 pTA in Europe

The European Commission has taken the initiative for fostering the dialogue between Science and Society and the involvement of citizens in S&T policymaking. The following EU projects have all tackled the idea of public dialogue on socio-ethical concerns about nanotechnology:

- Nanologue (2005–2006)
- NanoBio-RAISE (2006–2007)
- Nanoplat (2009)
- The framingNano Project (2008–2010)
- SNAP (2009–2010) (discussed briefly in Chapter 3)
- ObservatoryNANO (2008–2012)

The following brief discussions of each project suggest that there have been practical outcomes, ranging from tools for engagement, to methodologies for influencing nanotechnology stakeholders.

Nanologue, funded by Framework 6 of the EU, has as overarching objectives the establishment of a common understanding concerning social, ethical and legal aspects of nanotechnology applications and the facilitation of a Europe-wide dialogue among science, business and civil society about its benefits and potential impacts. Over 21 months of stakeholder consultation and scenario discussion, a

'NanoMeter', or online tool for assessing the risks of nanomaterials at the R&D stage, was developed (Nanologue 2006).

NanoBio-RAISE (Responsible Action on Issues in Society and Ethics) was intended to bring together nanobiotechnologists, ethicists and communication specialists to anticipate and discuss the societal and ethical issues likely to arise as nanobiotechnologies develop. In addition to various conferences and seminars, the project created a Democs (deliberative meeting of citizens) part card game, part policymaking tool that enables small groups of people to engage with complex public policy issues (Bruce 2007).

Nanoplat: The initial intention of the Nanoplat research project was to facilitate a form of deliberative process between the various players involved in defining, producing and commercialising that particular class of goods based on nanotechnology. It would thus have concrete power with which to influence the sector (Stø et al. 2011, p. 67). It aimed at the 'points of intersection between the sphere of production on the one hand and consumers on the other' (Nanoplat).

framingNano, launched in May 2008 with the aim of creating proposals for a workable governance platform, focused on analysing existing and ongoing regulatory processes, science-policy interfaces, researching risk assessment and governance in nanotechnologies as well as stakeholder consultation and information dissemination on governance. Recommending nano-specific regulation, the report noted 'uncertainties about public acceptance, resulting from a lack of transparency about EHS and ELSA issues' (Widmer et al. 2010).

The *ObservatoryNANO* project supports European policymakers through the provision of wide-ranging scientific and economic nanoanalysis and the assessment of ethical and societal aspects. It publishes an annual report and is working on an ethical toolkit for nanoapplication (Aitken et al. 2011).

In addition to the above-mentioned EU projects, the Netherlands, the UK and Germany offer other examples of public engagement. The Dutch 2009 'Nanopodium' project (which reported to parliament) concluded that:

- Citizens who had participated in the dialogues in 2009 and 2010 were in favour of responsibly implemented nanotechnology.
- Health-care applications attract most attention both for the potential benefits and for potential risks. Other applications of interest were food, personal care, security and privacy.
- The better the process of information, the more confident the citizens. Developing educational packages would be useful as young people are eager to learn about nanotechnology (Observatorio de Plastisco 2011).

The last point is of interest to policymakers and corporations concerned with consumer confidence in market terms. Thus, when the German NanoKommission (founded in 2006 by the German Federal Ministry for Environment, Nature Conservation and Nuclear Safety) [8] was set the task of discussing the opportunities

---

[8] In Germany, one institution may be regarded as the most prominent in the German TA landscape with regard to policymaking: the Institute of Technology Assessment and Systems Analysis (ITAS) at the Karlsruhe Institute of Technology (KIT).

and risks of nanotechnologies and formulating recommendations to the federal government, it casts its public net fairly wide in the Nanodialogues project, inviting representatives from the scientific world; the business community; environmental, consumer and women's associations; trade unions; churches; ministries; and authorities. The report and recommendations of the German NanoKommission (2010) noted that regarding corporations' risk management, guidelines needed to be more 'specific to business practice'.[9]

A 2009 UK reflection on the 2004 Royal Society Report argued that much of the clarity in the original report on the responsible development of nanotechnology had been lost and that more needed to be done in the field of public engagement. The report added that nanoparticles allegedly are being released into the environment and that, in the view of one collaborator on the updated report, there has been no 'meaningful change in regulatory practice or social engagement in the UK' (The Responsible Nano Forum 2009). The latter does not seem entirely justified, however, given the following UK nanodialogue initiatives that took place between 2004 and 2006 (plus the UK's involvement in the Dutch Democs project mentioned above):

- NanoJury UK, a citizen's jury (2005)
- The Small Talk programme (2004–2006), which sought to coordinate science communication-based dialogue activities
- The Nanodialogues project (2005–2006), a series of practical experiments to explore whether the public can meaningfully inform decision-making processes related to emerging technologies in four different institutional contexts
- The 'Moving Public Engagement Upstream' project (2004–2006), set up to examine the contribution of nanotechnology to sustainable development by developing socially and environmentally sensitive governance processes (Kearnes 2009)

One aspect of EU public engagement on nanotechnology issues not yet mentioned is that of school education. The EU 'NanoBioNet eV' provides vocational courses and training for teachers but has developed a multilingual (German, English and French) experimental kit ('the NanoSchoolBox') to teach school students about nanotechnology. Nanoethics, however, is not widely taught, in contrast to some Eastern counties, where we might find programs like the S&T 'impact on society' approach in the Korean school curriculum, or Japan's distribution of bioethics guidelines to schoolchildren. Taiwan has a nanoeducation program, 'K-12 Nanotechnology', started in 2002, that includes animations and graphic novels, as well as a high school lesson developed from Michael Crichton's nanonovel *Prey* (National Science Council 2009).

---

[9] In its final report presented in February 2011, the NanoKommission highlighted a broad set of recommendations for policy making with (among others) regard to risk and accompanying social research.

5.4   pTA in China

How does pTA in China differ from this energetic European pTA scene? Recent food scandals have catalysed S&T debates in China by highlighting the limitations of the policymaking system, leading to an erosion of public trust in science. This has been accompanied by increased calls for appropriate debates on policymaking processes that incorporate socio-ethical aspects along with economic ones. However, the problems of communication in such a society influenced by Confucianism mean that 'confronting others – especially elders – with truth or disagreement cannot be regarded as merely an exchange of information or of viewpoints but as an impolite and rude gesture' (Wang 1997). In Japan, for example, traditional forms of communication emphasise harmony, meaning peace and good relations with people in one's immediate surroundings (Takeichi 1997). In a communitarian society, strong debate is a difficult issue.

While expert TA has a good history in China, it tends to be dominated by natural scientists with little background in social issues. Information delivery, rather than debate, is more likely to be the aim of any survey.

Thus, the Chinese Institute of Physics disseminates research to the public through 'open houses'. 'Science popularisation' activities were carried out nationally in 2006 and 2010, with an upbeat message about new technology, particularly nanotechnology. Kearns, reporting on a 2009 UK-China workshop on the governance and regulation of nanotechnology, notes that in contrast to the UK's official commitment to upstream public debate, Chinese policy focuses more explicitly on the governance and coordination of research activities.

However, there is a movement towards greater public engagement with nanotechnology (Kearns 2009). When the Food Safety Law was revised in 2009, for example, in the legislative process to redraft the law, public opinion was solicited through government websites and newspapers, with over 11,000 comments being collected. With respect to nanotechnology, the annual Science Week held across China is also used to disseminate information about the sector while engaging public opinion (Jarvis and Richmond 2010).

In terms of nanotechnology, Chinese commentators and scholars acknowledge a lack of connection between research and universities in that nanodecisions are limited to specialists. Recently, the risk discourse has developed, focusing on the benefits of, and risks to, health, state security and ecology; the challenge to basic concepts such as 'health' and 'human'; the 'nanogap'; and the impact on privacy and human/consumer rights (Ying and Liao 2012).

While TA is familiar, pTA is a relatively new development. GM food, nanotechnology, stem cell technology and IT are still seen as major areas of attention for the scientific community, rather than the public. Yet there is a growing realisation among the scientific community that a wider approach, involving greater awareness of societal concerns, is required, potentially utilising pTA. A November 2011 workshop on scientific ethics worked to facilitate integration between the natural and the social sciences, suggesting that the Chinese scientific community realises that the social sciences have an integral role to play in technology evaluation and implementation. The argument in China currently appears to be that *nanoresearchers* bear the responsibility of proving the safety of nanotechnology (while developing such technology under the constraints of social-ethical norms), as well as of introducing nanotechnology to the public.

In China, the focus is on the responsibility of the scientist; in the EU, it is about government accountability to the public. The formulation of a code of conduct for scientists in both regions (predicted for 2012 in China, although not yet in evidence in 2013)[10] reveals similarities and differences in their respective approaches to nanotechnology development. This may change, since individual responsibility alone cannot guide S&T development, and public participation is increasingly being seen globally as integral to governmental decision-making. One might risk the prediction that this will ultimately become a governmental issue in China, since individual responsibility alone cannot guide S&T development (Decker and Li 2009).

pTA in China has so far been limited to a consensus conference on GM food in 2008, one with 25 participants. This conference was organised by the Chinese Academy of Sciences (CAS) and the S&T commission for Xicheng District, Beijing. Called 'Science and Community 2008 – GM foods', the aim was to recruit volunteers from varied socio-economic backgrounds; 38 applications were received, from which 20 public participants were chosen (plus 5 experts on GM foods and health, ethics and society, consumer choice and legislation). The final document report showed more uncertainty following the conference compared with responses beforehand about the relationship between the public and experts. Some participants began to 'modify their submissive position towards the experts' while remaining supportive towards the government. The primary result of the conference, it was concluded, was the clarification of the need for more public education. Du (2011) argues, however, that the '2008 consensus conference in Beijing indicated that public opinion has its own reference value, and the public also have some big picture thinking'. Zhou (2012) argues that there is an immature deliberative democracy in China, giving 'an optimistic prediction' for the future.

A 2012 report on public/government decision-making argued for a 'gradual movement towards policies and laws that…represent a more inclusive approach to policymaking at local and regional levels than has been traditionally employed' (International Association of Public Participation 2009). Interestingly, the same report, while noting that there is emerging public participation in the formulation of environmental regulation, quotes one interview on the 'weak' nature of community in China – communitarianism is more of a political than a civil society concept, meaning that public communities are seen as governmentally run rather than as demonstrations, or encouragement, of public responsibility.

Public participation is seen globally as increasingly integral to governmental decision-making, particularly given the economic effects of product boycotting, and

---

[10] According to GEST project information, due for updating in September 2013 at a project meeting in Beijing.

China might soon (and perhaps already is in terms of GM food) face this issue. The two regions may therefore be converging.

In conclusion, pTA does lead to certain outcomes, the most interesting of which is that of educating citizens, researchers and policymakers in a two-way or 'open' process that is consultative as well as informative. In terms of empowering the layperson by increasing his or her agency, it is seen as the foundation for a global ethics, as will be discussed below.

5.5   How pTA Works (and Why Habermas?)

There are two ways in which the efficiency of a pTA model can be analysed; one is logistical, asking what a pTA event involves, while the other focuses on the role of the agent in pTA. In both contexts, the views of Jürgen Habermas on discourse ethics offer some useful and practical guidelines that extend to the pTA model.

Rather than debating culturally complex and fraught questions of 'what is best', discourse ethics looks at the means by which such concepts might be debated and how such processes can be made more effective:

> Discourse ethics ... requires actual discussion and debate among those who may be affected by a norm or proposal and accepts the outcome as that which is morally correct, assuming of course that the debate was sound. Discourse ethics is, therefore, entirely procedural; it does not specify moral behaviors (sic) but only methods for agreeing upon them. In this, it would seem to have the potential for bringing about ongoing, practical resolutions of moral and ethical concerns. (Mingers and Walsham 2012)

Habermas's discourse ethics rests on two main ideas: universalisation and rational dialogue. Both are combined in the idea that practical consensus is achievable:

> Firstly, that contentions or utterances rest on particular validity claims that may be challenged and defended. For a normative ethical claim to be regarded as legitimate, it must be able to be justified in discourse. Arguers aim to construct cogent arguments that are convincing. The logical strength of such discourse depends on how well one has taken into account all the relevant information and possible objections, as argument implies counter-argument. Discourse is practical in that hypotheses are tested through argument.[11]

Habermas has been used in a variety of fields, in discussions on justice and democracy as well as on deliberative discourse, such as in Warren's (1993) argument that when individuals participate in democratic processes, they are likely to become more tolerant of differences, more attuned to reciprocity, better able to engage in moral discourse and judgment and more prone to examine their own preferences.

Habermas has been widely used in discussions of technology assessment, one example of which is Decker's (2002) analysis of 'rational technology assessment'.

---

[11] Practical discourse, Habermas (1991) writes, is 'a procedure for testing the validity of norms that are being proposed and hypothetically considered for adoption'.

He is less well known in the context of pTA. However, Hennen's (2012) recent work is an instance of usefully synthesising Habermasian theories, while Genus and Coles (2005) use both Habermas and Foucault in discussing CTA.

Reference is also made to Habermas's ideas in the work of Chambers (1994) on rural participation and in business ethics related to organisational communication and change (Stansbury 2009). Habermas appears in the work of Godin et al. (2007) on forensic health care, in Jacobson and Storey's (2004) study of the public communication of social change in Nepal and in Santos and Chess (2003) on how Army Restoration Advisory Boards facilitate Habermas's idealised conditions of speech as related to fairness.

Habermas has been popular in China since the 1980s, with the first book-length translation of his work appearing in 1989. His notion of the public sphere (*gonggon lingyu*, in Chinese) has proven particularly popular with Chinese intellectuals writing on deliberative democracy (Davies 2007).

How does Habermas's work add value to the idea of pTA?

5.5.1   The Logistics of a pTA Event

We may begin to answer this question by isolating the components and issues that a pTA/discourse event involves. This section, while chiefly describing and providing information about pTA, will also contain more discussion of the idea of education as a significant outcome of the pTA process.

The UK government has stated that 'properly targeted and sufficiently resourced public dialogue will be crucial in securing a future for nanotechnologies' (HM Government 2005). Yet achieving this is not necessarily easy. Thus, the Nanotech Engagement Group (NEG) in the UK (working with the Nanoscience Centre at Cambridge University and the Policy Studies Institute at East Anglia University) published a 2007 report on public engagement or, as the title of the report had it, 'democratic technologies', commenting on several of the above-listed initiatives. The seminal NEG report studied six UK upstream projects, identifying challenges such as:

- Creation of meaningful connections between public engagement and institutional decision-making
- Lack of understanding and appreciation in decision-making institutions and science communities of the different impacts and benefits that public engagement can deliver
- Lack of capacity and interest in public engagement within decision-making institutions and science communities
- A need to better distribute the benefits and impacts of public engagement (Gavelin et al. 2007)

5.5.2   Power and Pluralism

Perhaps the issue of power underlies the discourse/pTA event. 'Power' is here understood in the sense that the event is powerful if it has a democratic effect on policy. We are led to ask: (a) Does the inclusion of the layperson really have an effect on governmental action, and (b) is the exercise directed by government, scientists or institutions who have a predetermined agenda? In other words, who has the power to achieve positive action?

Point (a) has already been discussed. However, we can now refer to Habermas to see if he might add anything to this point. Habermas's (1982) notion of power includes both the negative idea of coercion from above and the positive idea of communicative power from below. His solution to the charge of ineffectiveness levelled against discourse ethics is that it needs to be complimented by a theory of socialisation that accounts for its institutionalisation:

> With discourse ethics as a guiding thread, we can indeed develop the formal idea of a society in which all potentially important decision-making processes are linked to institutionalized forms of discursive will-formation. (Habermas 1982)

Habermas reinforces the pTA model by reminding us that deliberative democracy has a legitimising function.

Responding to point (b), there is the issue of institutional location as well as that of audience, that is, we need to be cautious about such initiatives being firmly located in civil service departments, although government agencies and corporations might also be defined as 'locationally biased' (Gottweis 2000).[12] In other words, ownership – particularly funding – of the initiative or dialogue group requires scrutiny. This relates, in particular, to those conducting the meeting, for whom it is crucial that no claims of bias, disrespect or confrontational approach can be made. For the pTA/discourse process to be an exercise in deliberative democracy, the ownership of the information, even the questions, needs to be plural, arising both from the debate and being suggested to the debaters by the institutional or owning body. The criticism has been made that both pTA and discourse ethics reinforce the influence of the powerful, who are able to marshal more information and better-formed arguments than their less-informed, less-articulate and less-credible opposition. According to Habermas, the model requires openness and neutrality in the strictly defined sense of a free discourse without coercion. Only then can voices be heard fully and pluralistically.

Habermas (1993) suggests that the process of debate should aspire towards an 'ideal speech situation', in which everyone has equal rights and may question any assertion. Participants must come as close as possible to an ideal in which:

- All voices in any way relevant get a hearing.
- The best arguments available to us, given our present state of knowledge, are brought to bear.

---

[12] See Gottweis (2000) on the distribution of governance across both state and non-state actors and the creation of new spaces of participatory governance.

- Only the unforced force of the better argument determines the 'yes' and 'no' responses of the participants (Habermas 1993).

Ideally, no one capable of making a relevant contribution during the discourse process will have been excluded; participants will have an equal voice and be free to speak honestly in a free forum (i.e. one where no coercion is at work). Participants will recognise each other as having equal rights, will not lie and will not avoid critical understanding. As equal agents, inclusivity means equal and sincere inclusivity for them.

Discourse must take place in a forum containing many and varied representative viewpoints, so that discourse may be both pluralistic and dialogic. Thus, participants are encouraged to see the world from another's perspective and 'come to an understanding rather than merely a bargain' (Mingers and Walsham 2012). Habermas's model gives us reasons for why diversity is important, if a properly pluralistic and non-exclusionary debate is to take place rather than one in which the outcomes are already determined due to the homogenous nature of the invitees.[13]

Such pluralism also needs to have an impact on the decision-making outcomes. Irwin (2001) notes, in reference to the Public Consultation on Developments in the Biosciences held in the U in the 1990s, that it was really a government exercise, not a real consultation (it was instigated by the UK Department of Trade and Industry and conducted by Market and Opinion Research International Plc). He asks 'What happens when public opinion is opposed to government policy?'

A possible lack of clarity on hidden agendas like corporate influence also needs to be considered.

As the Dutch report on nanodialogues suggested, the framing of questions can be useful, particularly so as to differentiate between the risk issue and its social impact (Hanssen et al. 2008). Clear questions suggest clear objectives that ensure public engagement stays focused and expectations are managed. Flexibility will also be required though, as the process may lead to a redefinition of desired outputs. However, such flexibility can be subsumed within a single goal. In effect, the overall agenda will need to be set prior to directing the event but can be modified consensually through discourse.

5.5.3 Education?

With TA, the consensus of citizens relies on comprehending the issues, meaning there might be a need to look beyond any prevailing ideological framing of the debate. This translates the issue of power/agenda into one of representation/

---

[13] Habermas has been criticised for not fully appreciating issues of gender, race, ethnicity and sexuality or for understanding the pressures of historical contingency or other outside pressures that might introduce elements of coercion.

education. Or, in other words, the number of viewpoints and voices needs to be sufficiently broad and well informed to allow for deliberative democracy.

The issue therefore might be whether pTA requires too much knowledge from its laypersons. Gethmann (2002), looking at the issue of possibly overtaxing citizen competence, argues that this competence is not as highly regarded as scientists' expertise; however, one does not necessarily have to understand the science of nanomedicine to discuss issues of access and equity. pTA is societal rather than scientific.

pTA often implies a *process* of education, in that citizens, having familiarised themselves with a new science or technology, may contribute to an ongoing process of further participation. The dialogue between experts and laypersons in pTA leads to a mutual exchange of perceptions, as well as knowledge. The scientist's view that a shiny new technology is wonderful may encounter a layperson's view that such a shiny toy does not add value to life, and in fact she/he would rather do without it. After all, it is not only the public whose voice needs to be heard but also the scientist or TA practitioner, who needs to assume the role of the public intellectual – they 'go public' and consider wider issues than just those of the effects of technology, looking at ownership, control and social ends. Experts need to be prepared to listen to the public instead of regarding their role as that of making decisions on *behalf* of the public (Royal Society Report 2006).

This idea falls within the field of 'translational ethics'. As much as 'translational research attempts to connect the laboratory scientist's work to its implications for patient care, translational ethics focuses on bringing ethical scholarship into the sphere of personal and public action' (Cribb 2010). Schroeder (2012), in a draft paper at the Beijing GEST forum in November 2012, suggests summarising this concept as requiring:

- Ethical research that is compatible with knowledge generated in other disciplines
- Broad dissemination of results beyond one's own discipline
- Joining or leading attempts to implement results in the real world

Schroeder's model implies translation across disciplines as well as translation from the abstract to the practical. Yet 'translational' can also imply the translation not only of idea into reality but of translation across cultures. Farroni and Carter (2012) imply that there is movement beyond a compliance-based practice of ethics, intended to 'deepen the awareness and commitment to others' perspectives and values, respectful interactions, reflective problem-solving, and trust-building within, and between groups that engage in, as well as stand to benefit from translational research':

> Emerging from this model is the notion of translational ethics which requires a shifting of perspective and a realignment with the values and expectations of those who participate in research endeavors … the mere application of formal rules, codes, regulatory procedures, and ethical principles undermines and even ignores the moral content of the practice of translational science … a multidisciplinary, team-based approach in which ethical values and norms are generated, experienced and shared through relationships, dialogue, reflec-

tion, and support ... seeks to broaden the ethical competence of translational scientists, deepen understanding of the subjective experiences of research participants, and enlarge the capacity and sustainability of trust within research institutions and the communities they serve. (Farroni and Carter 2012)

Participants need to be open about their reasons for engagement and their expectations. Clarity of roles is significant as this makes it more likely that the process will achieve its objectives (Gavelin et al. 2007). In addition, the question 'what is in this for participants?' needs to be answered. As scientists often receive little incentive to engage with public dialogues, they may need support to discuss this with their affiliated institutions.

The task of the discourse/dialogue is to see what views can be 'universalised' and then 'consensualised'. In short, the education issue is not vital, as long as power relationships within the discourse/pTA event are neutralised – it should not be assumed that experts are in charge, 'that public knowledges are given the same status as scientific understandings' and that '*informative* (or information giving) and *consultative* (or information gathering) dimensions of participation' are balanced (Irwin 2001).

Decision-makers should ideally be involved in (invested in) the process early on. They should also commit, at the very least, to responding formally to its findings and recommendations. The involvement of policymakers would assist other participants to understand what is in fact feasible, just as policymakers could learn from lay opinion and from scientists about the complexities of governing science related to emerging technologies.

However, their presence can arguably distort the equality of the process, and so their role needs to be limited in terms of its input.

A final issue concerns how the findings are to be disseminated, so as to increase both public awareness and the level of public engagement itself while ensuring that any procedural mistakes are not replicated in future processes.

5.5.4 Other Logistical Issues: Size, Timeline and Methodology

A further issue in setting up pTA discussions is that of the size of the group. The NEG report refers to experiments in public engagement ranging from small focus groups of, for example, 13 people to much wider online participation. Whereas it is claimed that the interactive small group process allows for a depth of discussion often lost in large-scale public engagement, it nevertheless fulfils one objective in breaking down barriers, allowing people to interact on a more familiar basis. However, it carries the overriding problem of lacking a broad reach.

One country that has considerable nanotechnology stakeholder involvement has been Australia, yet the extent of the inclusion measured by group size has been small (Cormick 2010). On May 13, 2009, the Australian Government announced a 4-year National Enabling Technologies Strategy (NETS) to provide a framework for the responsible development of enabling technologies such as nanotechnology.

One key part of the strategy was the Enabling Technologies Public Awareness and Community Engagement (PACE) program that 'seeks to increase the public's awareness, knowledge and understanding of enabling technologies, including the risks and the benefits, to enable a more informed public debate' Russell 2013).

From 2007 to 2010, a number of events ranging from community presentations and educational events to more substantive fora were organised. Of the seven 'major' events, four were aimed at a better understanding and inclusion of public views and concerns. The forum on water involved only eight participants, however, and the one on bionics, nine (the reports of the fora did not include any meta-analysis of the process itself, focusing more on the positive engagement of participants) (Kyle and Dodds 2009).

In terms of timelines, public fora have ranged from one-off occasions to 2-year processes, including review meetings – the Nanoplat program is an example that introduced semi-directed online debates in order to review progress and evolve the debate.

A further issue is that of methodology. The question of how one organises the pTA process does not necessarily have a simple answer:[14]

> ...public participation and deliberative processes actually do not follow a given format. Rather, different forms of deliberative processes are used, from two hour card games on nanotechnologies to single evening events, focus group discussions of three hours' length, and processes running over half a year with 3 weekends for face to face contact and additional interaction in between these meetings. Accordingly, there are a variety of tools employed to stimulate interaction between participants, such as working groups, public hearings, plenary discussions, presentation plus question and answer sessions, scenario techniques, and card games. (Stø et al. 2011)

Methodology varies in trialled pTA fora, ranging from traditional meetings in which participants sit around a table to online methods. Considering the latter, the idea of groups that work on various topics and interact in person may be hard to ignore in terms of the benefits of direct personal contact for breaking down knowledge and cultural barriers. However, online tools can have a wider engagement utility; one instance was that of online gaming used in Citizen Science, a Bristol program aimed at engaging young people in discussions about the role of science and technology in society. The Democs conversation game is a similar idea, allowing participants to work through predeveloped policy decisions or to formulate one of their own.

In conclusion, the above issues suggest that the task of setting up an ideal pTA/ discourse event is complex, involving issues of representation, logistics, requiring interlocutors to be capable of rational compromise, and a democratic and inclusive process that will ideally affect or effect policy. The following questions can be extrapolated from reading Habermas's views on discourse along with those of com-

---

[14] Thomas Kilkauer (2010) suggests tens rules for setting up what he terms an 'ethics council' for corporations. These include inclusivity; communicative rationality; rules; the significance of time, place and layouts; moderators; and forms of agreement.

mentators on pTA and may hopefully serve as a useful checklist when setting up a pTA/discourse event:

1. What NELSI questions, apart from risk, have been framed for the discussion?
2. What are the desired outcomes of the process?
3. Who should be recruited for the process to ensure sufficiently diverse and minority-based viewpoints?
4. Who is excluded, and why?
5. How can certain social sectors be encouraged to participate, so as to avoid self-selection as the only criterion?
6. What logistical support is needed for the process (including budget requirements)?
7. Are there any hidden agendas?
8. How might participants best be prepared?
9. What is the optimal group size?
10. What methods, other than face-to-face discussion, are effective?
11. What is the most effective timescale? How many project phases are there?
12. Who conducts the process?
13. Who evaluates the process?
14. Who will be influenced by translating debate into policy?
15. What dissemination of information will result? What channels will be used for broad dissemination?

So much for general logistical issues. While the above offers a practical do-it-yourself guide to setting up your own nanoforum or nanodialogue, it has not, as yet, taken into account how or why this would be a useful method for *global* engagement. This requires us to look at the role of the agent in discourse.

It can be argued that the difficulty of achieving effective pTA lies in how dialogic the actual pTA/discourse forum is, in the sense of how external pressures (those of ownership, agenda setting, issues of power, etc.) and internal pressures (wishing one's voice to be heard) interrelate. If the process were truly Habermasian and dialogic, then it would be an *equal* one. Thus, the 'trick' of pTA is to ensure that the forum mode becomes dialogic, with participants able to affect changes to the conditions set up for them. How they accomplish this brings us to a discussion of the interlocutors/participants in the pTA process in the following chapter.

But there is a further aspect of duality. This other Habermasian facet of pTA is the duality of the *agent*, specifically, of his/her dual identity.

To introduce this concept, we might look broadly and briefly at the idea of global dialogue.

The award committee for the Global Dialogue Prize in 2009 nominated both Daryush Shayegan and Mohammad Khatami as joint winners for their work in developing and promoting the concept of a dialogue among cultures and civilisations as a new paradigm for international relations that includes cultural subjectivity, i.e. that operates as a pluralist model that does not ignore cultural relativism.

Khatami transposed the ideas of Daryush Shayegan (Director of the Iranian Centre for the Study of Civilizations, 1976–1979), particularly his notion of intercultural contact as bound to modify the ideas and values in any of the participating cultural systems. Shayegan's dialogue paradigm promotes active listening and adopts some features of the virtue ethics framework. Arguing that politics should be tied closely to ethics, he advocates moral virtues and psychological dispositions such as modesty, commitment and involvement, as well as sympathy and affection, in a 'genuine effort to understand others without the desire to vanquish them'. Khatami (1999) adopts what seems rather a Habermasian idea of dialogue, involving a rational reconsideration of faith and arguing that the latter should be open to change, referring to 'the rational maturity of human beings'.

Khatami was instrumental in promoting the UN 2001 'Year of the Dialogue Amongst Civilisations', the report of which was published as *Crossing the Divide: Dialogue among Civilizations and the Global Agenda for Dialogue among Civilizations*. Speaking at a Symposium on the Dialogue of Civilisations in 2000, Khatami offered the following definition of dialogue:

> Dialogue entails a clear and precise understanding of the world's cultural geography. It means taking a critical look at the 'self' and the 'other'. It means paying attention to heritage as well as being serious about learning new experiences. Dialogue concerns humanity's needs today and tomorrow. Therefore, opening a new door towards understanding global realities, and finding new viewpoints in the East and West, are prerequisites of real dialogue between civilisations and cultures. The basic question is, How can one find a common landscape to view, a common message to hear and a common language to speak? It is not possible to engage in dialogue with deaf ears and an unfamiliar language. One must enter dialogue on the basis of Eastern and Western values… Westerner and Easterner can remain different and complementary in the parallel realms of cultures and maintain their emotional affiliation to their origins. (Khatami 2001)

The above suggests that international dialogue rests on notions of self and other, i.e. on a duality of culturally learned values as well as receptivity to the values, thoughts and feelings of the other. Shayegan suggested this in his acceptance speech for the Global Dialogue Prize:

> we live in a fragmented world of broken ontologies, a world in which the concept of interconnectiveness has been substituted for the old metaphysical foundations. And this interconnectiveness manifests itself at all levels of culture, knowledge and science: multiculturalism, plural identities, World Wide Web, holistic science. We live in a world of hybrid cultures where all levels of consciousness overlap each other. If assumed with lucidity and without resentment this new mosaic configuration can enrich us, extending the registers of knowledge, enlarging the range of feelings. (Shayegan 2010)

Educating themselves in the ways of the other, in strange ontologies, agents have to adopt a 'dual self', a key concept for the remainder of this study. The argument is that the global agent is a virtuous citizen who can adopt a dual identity that will allow consensus. Habermas's interlocutor or agent must be capable of reason and be capable of achieving some form of cultural compromise that might allow for universalisable or reconciled ideas. Participants accept good arguments self-critically (Bohler 1993).

Does such a paragon exist? One might add, briefly, that Habermas's views are not as utopian as is often claimed. Discourse consensus will be based on an evolving compromise (e.g. a majority vote). Decisions remain open to further improvements that make the discursive process dynamic, for 'provisionally justified views might have to be revised in the light of new information and arguments' (Habermas 1998).

These topics will be discussed in the following chapter.

In conclusion, one can summarise that Habermas's model requires rational and polyphonic agency within dialogue. In terms of dialogue, the interlocutor attempts mediation between his/her own motives, needs and virtues and the procedural requirements of the discourse, adapting his or her views according to the procedural imperative. This requires self-awareness or 'emancipatory knowledge', as well as an intersubjective approach to ethics.

The latter is the basis for the theory of 'dual identity' on which this study predicates the feasibility of global ethics. This requires a discussion of identity in terms of the agent, as will be seen in the next chapter.

References

Abels G (2007) Citizen involvement in public policy-making: does it improve democratic legitimacy and accountability? The case of pTA. Interdiscip Inf Sci 13(1):103–116

Abels G (2010) Participatory technology assessment and the institutional void. Investigating democratic theory and representative politics. In: Bora A, Hausendorf H (eds) Democratic transgressions of law: governing technology through public participation. Brill, Leiden, pp 239–268

Aitken RJ, Ross BL, Peters SAK, Geertsma RE, Bleeker E, Wijnhoven SWP, Toufektsian MC, Nowack B (2011) ObservatoryNANO. http://www.triwu.it/documents/10157/7fc31ea5-8b80-4b69-9224-469b7ff6bafb. Accessed 27 Dec 2014

Arras J (2010) Theory and bioethics. In: Zalta EN (ed) The Stanford encyclopedia of philosophy, Summer edn. http://plato.stanford.edu/archives/sum2010/entries/theory-bioethics/. Accessed 27 Dec 2014

Benhabib S (1993) Communicative ethics and current controversies in practical philosophy. In: Benhabib S, Dallmayr F (eds) The communicative ethics controversy. MIT Press, Cambridge, pp 330–369

Blum A (2012) Public engagement and risk governance of nanotechnologies – revolution or illusion? Dissertation, University of Wales

Bohler D (1993) Transcendental pragmatics and critical morality. In: Benhabib S, Dallmyr RM (eds) The communicative ethics controversy. MIT Press, Cambridge, pp 111–150

Bora A, Hausendorf H (2010) Governing technology through public participation. In: Democratic transgressions of law: governing technology through public participation. Brill, Leiden, pp 1–20

Bostrom N (2007) Technological revolutions and the problem of prediction. In: Lin P, Allhoff F, Moor J, Weckert J (eds) Nanoethics: the ethical and social implications of nanotechnology. Wiley, Hoboken, pp 101–118

Brom F, van Est R (2011) Risk and technology assessment. Available via Rathenau Institute. http://www.rathenau.nl/en/publications/publication/risk-and-technology-assessment.html. Accessed 27 Dec 2014

Bruce D (2007) Engaging citizens on nanobiotechnology using the DEMOCS game: interim report on the DEMOCS games on nanobiotechnology played in the UK and the Netherlands. http://ethics.iit.edu/NanoEthicsBank/node/1751. Accessed 27 Dec 2014

Burmeister O, Weckert J, Williamson K (2011) Seniors extend understanding of what constitutes universal values. J Inf Commun Ethics Soc 9(4):238–252

Chambers R (1994) Participatory rural appraisal (PRA): analysis of experience. World Dev 22(9):1253–1268

Collins H, Evans R (2007) Rethinking expertise. University of Chicago Press, Chicago

Cormick C (2010) The challenges of community engagement. NanoEthics 4:229–231

Cribb A (2010) Translational ethics? The theory-practice gap in medical ethics. J Med Ethics 36:207–210

Crocker DA (2008) Ethics of global development. Agency, capability and deliberative democracy. Cambridge University Press, Cambridge

Davies G (2007) Habermas in China: theory as catalyst. China J 57(January):61–85

Observatorio de Plastico (2011) Dutch nanodialogue concluded. http://www.observatorioplastico. com/detalle_noticia.php?no_id=150322&seccion=mercado&id_categoria=803. Accessed 27 Dec 2014

Decker M (2002) Interdisciplinarity in technology assessment: implementation and its chances and limits. Springer, New York

Decker M, Li Z (2009) Dealing with nanoparticles: a comparison between Chinese and European approaches to nanotechnology. In: Ladikas M (ed) Embedding society in science & technology policy. European and Chinese perspectives. European Commission, Brussels, pp 91–123

Delanty G (2000) Citizenship in a global age. Society, culture, politics. Open University Press, Buckingham

Dissanayake W (1996) Introduction. In: Dissanayake W (ed) Narratives of agency. Self-making in China, India, and Japan. University of Minnesota Press, Minneapolis, pp ix–xxi

Dryzek J, Tucker A (2008) Deliberative innovation to different effect: consensus conferences in Denmark, France and the United States. Public Adm Rev 68(5):864–876

Du P (2011) The practice of TA in China. Paper presented at NCSTE. Chinese Academy of Science, Bejiing, 10 November 2011

European Commission (2010) Deepening ethical engagement and participation in emerging Nanotechnologies. http://cordis.europa.eu/project/rcn/84695_en.html. Accessed 27 Dec 2014

European Year of Citizens 2013 (2013) http://ey2013-alliance.eu/. Accessed 27 Dec 2014

Fang Y (2011) Dialogue a sign of increasingly sincere ties. China Daily, 16 September. http://usa. chinadaily.com.cn/opinion/2011-09/16/content_13723556.htm. Accessed 27 Dec 2014

Farroni J, Carter M (2012) Translational ethics: an Engaged humanities approach. Paper presented at the Association for Clinical Research Training Conference, Washington, DC, 18–20 April 2012

Fox W (2006) A theory of general ethics. MIT Press, Cambridge

Friedman HB (1997) Human values and the design of computer technology. Cambridge University Press, New York

Fuller S (2000) The governance of science: ideology and the future of the open society. Open University Press, Buckingham

Gaskell G, Thompson P, Allum N (2002) Worlds apart? Public opinion in Europe and the USA. In: Bauer MW, Gaskell G (eds) Biotechnology – the making of a global controversy. Cambridge University Press, Cambridge

Gavelin K, Wilson R, Doubleday R (2007) The final report of the nanotechnology engagement group (NEG). Available via Involve. http://www.involve.org.uk/wp-content/uploads/2011/03/ Democratic-Technologies.pdf. Accessed 27 Dec 2014

Genus A, Coles A (2005) On constructive technology assessment and limitations on public participation in technology assessment. Technol Assess Strateg Manag 17(4):433–443

Gethmann CF (2002) Participatory technology assessment: some critical questions. Poiesis Praxis Int J Technol Assess Ethics Sci 1(2):151–159

Gewirth A (1982) Human rights: essays on justification and application. University of Chicago Press, Chicago

Gewirth A (1984) The epistemology of human rights. Soc Philos Policy 1(2 Spring):14–17

Godin P, Davies J, Heyman H, Reynolds L, Simpson A, Floyd M (2007) Opening communicative space: a Habermasian understanding of a user-led participatory research project. J Forensic Psychiatr Psychol 18(4):452–469

Gottweis H (2000) Participation and the new governance of life. BioSocieties 3:265–286

Gov.Uk. Horizon Scanning Programme Team. http://www.bis.gov.uk/foresight/our-work/horizon-scanning-centre. Accessed 27 Dec 2014

HM Government (2005) Response to the Royal Society and Royal Academy of Engineering report: nanoscience and nanotechnologies: opportunities and uncertainties. http://news.bbc.co.uk/nol/shared/bsp/hi/pdfs/25_02_05nanotech_final.pdf. Accessed 27 Dec 2014

Guston D (2011) Participating despite questions: toward a more confident participatory technology assessment. Sci Eng Ethics 17(4):691–697

Habermas J (1982) A reply to my critics. In: Thompson JB, Held D (eds) Habermas: critical debates. MIT Press, Cambridge, pp 219–283

Habermas J (1991) Moral consciousness and communicative action. MIT Press, Cambridge

Habermas J (1993) Justification and application: remarks on discourse ethics. MIT Press, Cambridge

Habermas J (1996) Three normative models of democracy. In: Benhabib S (ed) Democracy and difference. Princeton University Press, Princeton, pp 21–30

Habermas J (1998) Between facts and norms. MIT Press, Cambridge

Hanssen L, Walhout B, van Est R (2008) Ten lessons for a nanodialogue. The Dutch debate about nanotechnology thus far. Rathenau Institute, The Hague

Hennen L (1999) Participatory technology assessment: a response to technical modernity? Sci Public Policy 26(5):303–312

Hennen L (2012) Why do we still need participatory technology assessment? Poesis Prax 9:27–41

Hodge GA, Bowman DM (2007) Engaging in small talk: nanotechnology policy and dialogue processes in the UK and Australia. Aust J Publ Adm 2(66):223–237

Horst M (2010) Collective closure? Public debate as the solution to controversies about science and technology. Acta Sociol 53(3):195–211

Horvat M, Lundin N (2008) Review of the Science & Technology Cooperation between the European community and the government of the People's Republic of China. http://ec.europa.eu/research/iscp/pdf/policy/st_agreement_eu_china2008.pdf. Accessed 27 Dec 2014

Hutchings K (2010) Global ethics: an introduction. Polity Press, Cambridge

International Association of Public Participation (2009) Painting the landscape. A cross-cultural exploration of public – government decision making. http://c.ymcdn.com/sites/www.iap2.org/resource/resmgr/imported/Kettering_FINALExecutiveSummaryReport.pdf. Accessed 27 Dec 2014

Irwin A (2001) Constructing the scientific citizen: science and democracy in the biosciences. Public Underst Sci 10:1–18

Jacobson T, Storey J (2004) Development communication and participation: applying Habermas to a case study of population programs in Nepal. Commun Theory 14(2):99–121

Jarvis D, Richmond N (2010) Mapping emerging nanotechnology policies and regulations: the People's Republic of China. http://www.spp.nus.edu.sg/docs/wp/2010/wp1005.pdf. Accessed 27 Dec 2014

Jones R (2009) Public engagement and nanotechnology – the UK experience. Soft Machines. http://www.softmachines.org/wordpress?p-443. Accessed 27 Dec 2014

Joss S (1999) Public participation in science and technology policy- and decision-making – ephemeral phenomenon or lasting change? Sci Public Policy 26(5):290–293

Joss S (2002) Toward the public sphere – reflections on the development of participatory technology assessment. Bull Sci Technol Soc 22(3):220–231

Jotterand F (2006) The politicization of science and technology: its implications for nanotechnology. J Law Med Ethics 34(4):658–666

Kaiser M, Kurath M, Maasen S, Rehmann-Sutter C (eds) (2009) Governing future technologies: nanotechnology and the rise of an assessment regime. Springer, Dordrecht

Kearnes M, Macnaghten P, Wilsdon J (2006) Governing at the nanoscale. People, policies and emerging technologies. http://www.demos.co.uk/files/governingatthenanoscale.pdf?1240939425. Accessed 27 Dec 2014

Kearns M (2009) A UK-China workshop: governance and regulation of nanotechnology: the role of the social sciences and humanities. http://www.docstoc.com/docs/17707152/ UK-Chinaworkshop-Governance-and-Regulation-of-Nanotechnology. Accessed 27 Dec 2014

Khatami M (1999) Dialogue and the new millennium, address to the annual session of the United Nations Educational, Scientific and Cultural Organization (UNESCO) on 29 October 1999. http://en.rafed.net/index.php?option=com_content&view=article&id=9557:dialogue-and-the-new-millennium&catid=81:miscellaneous&Itemid=846. Accessed 27 Dec 2014

Khatami M (2001) Symposium: Islam, Iran and the dialogue of civilisations. Glob Dialogue 3(1) (Winter). http://www.worlddialogue.org/content.php?id=125. Accessed 27 Dec 2014

Kilkauer T (2010) Critical management ethics. Palgrave Macmillan, Basingstoke

Korsgaard C, Cohen GA, Geuss R, Nagel T, Williams B (1996) The sources of normativity. Cambridge University Press, Cambridge

Kyle R, Dodds S (2009) Avoiding empty rhetoric: engaging publics in debates about nanotechnologies. Sci Eng Ethics 15:81–96

Ladikas M (2009) Introduction. In: Ladikas M (ed) Embedding society in science & technology policy. European and Chinese perspectives. European Commission, Brussels, pp 13–14

Ladikas M, Schroeder D (2005) Too early for global ethics? Camb Q Health Care Ethics 14(4):404–410

Levidow L (2009) Democratizing agri-biotechnology? European public participation in agbiotech assessment. Comp Sociol 8(4):541–564

Loeber A, Griessler E, Versteeg W (2011) Stop looking up the ladder: analyzing the impact of participatory technology assessment from a process perspective. Sci Public Policy 38(8):599–608

Macnaghten PM, Kearnes MB, Wynne B (2005) Nanotechnology, governance and public deliberation: what role for the social sciences? Sci Commun 27(2):268–287

McGiffen S (2005) Biotechnology. Corporate power versus the public interest. Pluto Press, London/Ann Arbor

Mingers J, Walsham G (2012) Toward ethical information systems: the contribution of discourse ethics. MIS Q 34(4):833–854

Nanologue (2006) Opinions on the ethical, legal and social aspects of nanotechnologies – results from a consultation with representatives from research, Business and Civil Society. http://nano-tech.law.asu.edu/Documents/2009/09/NanologueWP34FinalPublic_233_8071.pdf. Accessed 27 Dec 2014

National Science Council, Taiwan (2009) National science and technology development plan 2009–2012. http://www.most.gov.tw/public/Attachment/91214167571.PDF. Accessed 27 Dec 2014

Nelkin D (1975) The political impact of technical expertise. Soc Stud Sci 5:35–54

Nowotny H, Scott P, Gibbons M (2001) Re-thinking science: knowledge and the public in an age of uncertainty. Polity, Cambridge

Nussbaum M (1988) Nature, functioning and capability: Aristotle on political distribution. Oxf Stud Anc Philos 6:145–184

Nussbaum M (2000) Women and human development: the capabilities approach. Cambridge University Press, Cambridge

O'Brien DM, Marchand DA (1982) The politics of technology assessment. Institutions, processes and policy disputes. D.C. Heath & Co, Lexington

Pidgeon N, Hayden TR (2007) Opening up nanotechnology dialogue with the publics – risk communication or 'upstream engagement'? Health Risk Soc 9(2):191–210

Pogge TW (1999) Human flourishing and universal justice. In: Paul EF, Miller FD Jr, Paul J (eds) Human flourishing, vol 16. Cambridge University Press, Cambridge

Porter AL, Rossini FA, Carpenter SR (1980) A guidebook for technology, assessment and impact analysis. Elsevier North Holland, New York

Rawls J (1971) A theory of justice. Harvard University Press, Cambridge, MA

Rocco M, Bainbridge WS (eds) (2001) Societal implications of nanoscience and nanotechnology. Kluwer, Boston

Royal Society (2006) Royal Society report: science communication: excellence in science Royal Society. http://royalsociety.org/uploadedFiles/Royal_Society_Content/Influencing_Policy/Themes_and_Projects/Themes/Governance/Final_Report_-_on_website_-_and_amended_by_SK_no_navigation.pdf. Accessed 27 Dec 2014

Russell AW (2013). Improving legitimacy in nanotechnology policy development through stakeholder and community engagement: forging new pathways. http://onlinelibrary.wiley.com/doi/10.1111/ropr.12037/abstract;jsessionid=669F8AE4B020243A60978C92952E3FAC.f04t01?deniedAccessCustomisedMessage=&userIsAuthenticated=false. Accessed 28 Dec 2014

Sandler R (2007) Character and environment. Columbia University Press, New York

Santos S, Chess C (2003) Evaluating citizen advisory boards: the importance of theory and participant-based criteria and practical implications. Risk Anal Int J 23(2):269–279

Satterfield T, Kandlikar M, Beaudrie C, Conti J, Herr Harthorn B (2009) Anticipating the perceived risk on nanotechnologies. Nat Nanotechnol 4(November):752–758

Schummer J (2004) Societal and ethical implications of nanotechnology: meanings, interest groups and social dynamics. Techno 8:56–87

Sciencewise Report (2011) International comparison of public dialogue on science and technology. http://www.sciencewise-erc.org.uk/cms/assets/Uploads/Publications/International-Comparison-of-Public-Dialogue.pdf. Accessed 27 Dec 2014

Sclove R (2010) Reinventing technology assessment: a 21st century model. http://www.loka.org/documents/reinventingtechnologyassessment1.pdf. Accessed 27 Dec 2014

Sen A (1977) Rational fools: a critique of the behavioral foundations of economic theory. Philos Publ Aff 6(4):317–344

Sen A (1980) Equality of what? In: McMurrin S (ed) The Tanner lectures on human values. Cambridge University Press, Cambridge, pp 197–220

Sen A (1993) Capability and well-being. In: Nussbaum M, Sen A (eds) The quality of Life. Oxford University Press, Oxford, pp 30–53

Sen A (1999) Development as freedom. Oxford University Press, Oxford

Shayegan D (2010) The dialogue of civilisations. http://www.globaldialogueprize.org/page.php?idMenu=1&idSub=1&idMain=74. Accessed 27 Dec 2014

Siegrist M, Keller C, Kastenholz H, Frey S, Wiek A (2007) Laypeople's and experts' perceptions of nanotechnology hazards. Risk Anal 27(1):59–69

Silverman H (2010 Participatory technology assessment. People Place 1(3). http://www.people-andplace.net/perspectives/2010/9/13/participatory_technology_assessment. Accessed 27 Dec 2014

Slob M, Raeymaekers P, Rondia K (2005) Meeting of minds. Food for thought and debate on brain science. http://www.danacentre.org.uk/documents/pdf/food_for_thought_and_debate_on_brain_science.pdf. Accessed 27 Dec 2014

Spence EH (2007) Positive rights and the cosmopolitan community: a rights-centred foundation for global ethics. J Glob Ethics 3(2):181–202

Stansbury J (2009) Reasoned moral agreement: applying discourse ethics within organizations. Bus Ethics Q 19(1):33–56

Stø E, Scholl G, Jègou F, Strandbakken P (2011) The future of deliberative processes on nanotechnology. Zeppelin, Freidrichshafen, pp 53–80

Stockholm International Peace Research Institute (2015) EU common position. http://www.sipri.org/research/armaments/transfers/controlling/eu_common_position. Accessed 20 Dec 2015

Takeichi H (1997) Japanese-style communication in a new global age. In: Christians C, Traber M (eds) Communication ethics and universal values. Sage, Thousand Oaks, pp 244–248

The Responsible Nano Forum (2009) A beacon or just a landmark? Reflections on the 2004 Royal Society/Royal Academy of Engineering report: nanoscience and nanotechnologies: opportunities and uncertainties. http://www.nanotechproject.org/process/assets/files/8273/beacon_or_landmark_report_-_rnf.pdf. Accessed 27 Dec 2014

van Est R (2008) Ten lessons for a nanodialogue. About being serious and having some serious fun. http://www.oecd.org/sti/nano/42326543.pdf. Accessed 27 Dec 2014 van Est R (2011) Keeping the dream alive: what ELSI research might learn from parliamentary technology assessment. In: Cozzens S, Wetmore J (eds) Nanotechnology and the challenges of equity, equality and development, Yearbook of nanotechnology in society. Springer, Dordrecht, pp 409–421

van Est R, Walhout B, Rerimassie V, Stemerding D, Hanssen L (2012) Governance of nanotechnology in the Netherlands – informing and engaging in different social spheres. Int J Emerg Technol Soc 10:6–26

Venn C (2002) Altered states: post-enlightenment cosmopolitanism and transmodern socialities. Theory Cult Soc 19(1–2):65–80

Walsh B (2009) Environmentally beneficial nanotechnologies: barriers & opportunities. DEFRA, London

Wang G (1997) Communication ethics in a changing Chinese society: the case of Taiwan. In: Christians C, Traber M (eds) Communication ethics and universal values. Sage, Thousand Oaks, pp 225–258

Warren M (1993) Can participatory democracy produce better selves? Psychological dimensions of Habermas' discursive model of democracy. Polit Psychol 14(2):209–234

White SK (1982) On the normative structure of action: Gewirth and Habermas. Rev Polit 34(2):282–301

Widmer M, Meili C, Mantovani E, Porcari A (2010) FramingNano. Governance in nanoscience and nanotechnology. http://www.nanotec.it/public/wp-content/uploads/2014/04/FramingNano_GovernancePlatformFinal_Report.pdf. Accessed 27 Dec 2014

Ying M, Liao M (2012) Nano-technology development in China and the related ethic discussions. Paper presented at the GEST roundtable, Bejiing, 5 September 2012

Zhou W (2012) In search of deliberative democracy in China. J Publ Deliber 8(1):1–17

Chapter 6
The Virtuous Discourse Agent

> *All virtue is summed up in dealing justly – Aristotle (Thomson 1953)*
>
> *On the whole, human beings want to be good, but not too good, and not quite all the time – Orwell (Orwell and Gessen 2008)*
>
> *Each must be able to recognise him- or herself in all that wears a human face (Habermas 1992).*

Abstract This chapter examines the idea of 'global virtues', looking at Aristotelian, or Western-centric virtues, Habermasian 'virtues' or skills', and Chinese jen or ren ethics, encapsulating such ideas as non-maleficence, respect for others, 'human heartedness', benevolence, love, compassion, kindness, and humanity.

Keywords Aristotle • Virtue ethics • Ren ethics • Ubuntu • Habermas

Consequentialism and deontology are less useful approaches to take to pTA than virtue ethics.[1] Significant activity or agency is the key to this model of discourse ethics/pTA.

Williams (1985) and MacIntyre (1981) have suggested that the theories of consequentialism and deontology impose a universalist framework, ignoring the role of personal commitment, as well as that of localised community values, or reciprocity, thus Whetstone's (2001) argument that virtue ethics is required. Kirchengast (2011) suggests that virtue ethics is particularly useful for global ethics:

> Virtue ethics has much to offer…because with its naturalistic idea of humans as rational beings whose natural way of life is communal, it might be of more appeal to proponents of other cultures than, for example, a moral theory grounded on an overly individualistic notion of humanity. Saying that 'virtue ethics might have more appeal' to members of non-

---

[1] The arguments usually levelled against virtue ethics as an approach are, broadly speaking, that it is (a) too subjectively self-absorbed and too narrow, concepts such as global justice not being well developed within the theory; (b) lacking normative principles, and (c) impotent. See Hursthouse (1999), Meilaender (1984), and Walker and Ivanhoe (2007).

The 'justification problem' in virtue ethics (of how we justify or agree on which traits are virtues) may need to be supplemented by other ideas such as coherentism or ethical naturalism; the 'yardstick' of 'flourishing' used in virtue ethics is seen as inadequate. See Conley (1988).

Western cultures…(suggests) that its method… is one that is open to proposals differing
from traditions other than Western philosophical positions… Virtue ethics'…strength lies,
among other things, in its conviction that all human beings share the same basic nature and
accordingly have the same basic needs, so that a common set of values is in principle
possible.

While agents of discourse are required to discuss the consequences of new technology, and to examine the morality of the technological activity, the main aim for
such agents is to achieve rational dialogue. To do this, they need to possess those
virtues that impel them to fully enter into intersubjective dialogue.

Virtue ethics is associated most closely in the West with Aristotle, whose theory
of potentiality implies that all matter in the universe, people included, is driven
towards perfectibility or, rather, towards the fulfilment of that potential. This is both
the path and the goal of the person living the truly virtuous, happy life. How does
Habermas 'update' the theory of the virtuous human, so that the participant in discourse ethics might enter dialogue with the 'right attitudes'? And how universal
might these virtues be?

Habermas (1998) has never explicated his concept of virtue and indeed often
criticises the concept, believing that it plays down the importance of individual
qualities, and presupposes a shared conception of the good life. Since Habermasian
discourse eschews hegemonic bias, any shared preconception should be avoided in
favour of neutral, practical procedures. Habermas is far less concerned with – in
fact, abhors – telling his readers how to achieve the good life than with establishing
conditions for practical dialogue and debate. Aristotle's *telos* of virtue – the 'good'
towards which humanity strives – is based around the notion of *eudaemonia*, the
prosperous, noble, and virtuous happiness of the Athenian gentleman. The
Habermasian *telos* is that consensual aim of communication:

For Habermas, this *telos* is the end of coercion and the attainment of autonomy through
reason, the end of alienation through a consensual harmony of interests, and the end of
injustice and poverty through the rational administration of justice. (Braaten 1991)

Communicative action is an inherently consensual form of social coordination,
in which actors mobilise the potential for rationality given by ordinary language and
its *telos* of rationally motivated agreement. Although it attempts to externalise virtue into procedures as much as possible, discourse ethics presupposes that the participants possess a virtuous attitude – why else would they wish to engage in
discourse? Habermas's system is a procedural ethics not linked to any substantive
values but one that nonetheless implies certain virtues, demonstrated through discourse skills, on the part of the discourse agent.

Perhaps agents enter discourse for the purely egoistic reason of demonstrating
their rational skills at interlocution, though that would be a fairly empty exercise.
Like Aristotle's Athenian gentleman, Habermas's discourse agent must be civic
minded and committed to the 'greater good' of a rational discourse process. In
identifying moral behaviour with discourse, Habermas implies a universal or perhaps primary virtue – that of willingness to communicate rationally with the intent
to reach agreement. However, Habermas does not call this a virtue so much as a
'species ethic'.

6.1   The Species Ethic as Universal?

Speech is the mirror of the soul; as a man speaks, so he is – Publilius Syrus (Lyman 1856)

To be human entails having the great (universalist) desire to communicate one's views to another human being (whether opposing or not). Habermas's view of what it is to be human suggests that an essential part of the definition requires us to be members of a moral, language-based community.[2]

In his *The Future of Human Nature* (2003), Habermas suggests that reason, which leads to moral consensus, is sustained by a 'species ethic' ('Gattungsethik'), or prior ethical self-understanding of the species shared by all moral persons.

We can disentangle this thought as follows: whenever agents use language to coordinate their actions, they enter into certain commitments to justify their words on the basis of good reasons and so agree to apply reasoning to their words. These commitments are not merely procedural; they also have a moral status, being universally applicable to agents (Finlayson 2005). Thus, to be human is to be capable of morality and to actualise that capability within society, through the intersubjectivity provided by being a language user. The species ethic presupposes that we approach others as equals and that we want to be moral, that is, that we want to live with those others in a society that is so constituted as to protect and nourish ours, as well as their ethical decisions about our own lives and how we live them. In other words, our social arrangements lead us to prioritise the autonomy or dignity of the individual and therefore to draw limits on the impositions that can be made on others.

Habermas's (2003) idea leads him to approve of Rawls's view of the 'just society' as one in which individuals choose how it is that they want to 'spend the time they have for living', guaranteeing equal freedom to develop an 'ethical self-understanding, so as to realise a personal conception of the good life according to one's own abilities and choices'. These ideas, of course, hint at the Western-centric nature of Habermasian thought – namely, when looking at issues of justice and autonomy, beloved by Western commentators, though rather less so by Eastern ones, in that they often imply the rights of the individual as paramount. In a just society, the individual may have the autonomy that implies his/her right to justice; but is the right of the society as a whole to justice more important than that of the individual?[3]

With his concept of the species ethic, Habermas makes various assumptions that may be neither feasible nor likely to be universal, namely, that the human being is:

1. A social animal…

---

[2] The reader will realise that this topic introduces a larger one, favoured by evolutionary psychologists, but outside of the limits of this study.

[3] Given the nanotechnology focus of this study, it is worth noting that in his writings on the species ethic, Habermas has a somewhat conservative view of change. Genetic *enhancements*, for example, are described as a form of (albeit positive) eugenics and so should be forbidden, in his view.

2. Aware of his/her civic duty to participate in public debate, to work towards consensus (Cortina 2008; Apel 2008)
3. Willing to work *rationally* towards consensus and so to obey the rules of discourse
4. Sharing a sense of autonomy and justice with others so that he/she enters debate 'equally' with others – public 'morality', by which Habermas really seems to mean justice, is based on one's commitment to such equality

These are all weighty issues. Taking the perhaps idealistic view that the human being strives to be moral instead of following his/her own strategic aims, we must ask from where comes our belief that others are worth helping or treating with care (Taylor 1989). Whence comes our desire to be social animals aware of our civic duty? Habermas argues that we cannot refuse to participate in society, as this would mean that the 'refuser' effectively chooses not to recognise herself or himself as a member of a community and is therefore not counted as a member of that community. Therefore, we are all inevitably civic minded, with discourse the default method for settling social disputes, unless we choose to live in splendid social isolation, communicating with nobody. Loners might state that they are beyond the bounds of the species ethic; yet misanthropy implies a *dislike* of humanity which is not the same as being utterly 'other'. The misanthropic voice is arguably part of the human debate, despite its wish to reject it.

If that sounds slightly dubious in logical terms, then Habermas has another answer in terms of a grounded context for one's subjectivity. Even if one refuses the *ur*-context or macro-context of the species ethic, one has a background micro-context relating to one's family, education, and so forth that still renders one 'human' and thus a social animal. It would be a rare person who was raised by wolves, without social or cultural contexts grounded in and explicated through language.

In addition, Habermas would claim that the moral dimensions of language lie in the very act of speaking with another. Entering discourse not only recognises our civic mindedness, it *develops* it. Our fundamental human interest, as Habermas terms it, is in communication; and through developing this interest, we recognise and develop ourselves. In short, he makes a vital connection between speaking and being a social person. The species ethic should be seen as something that equips us for dialogue; if we refuse to be heard by engaging in such dialogue, we are failing another of Habermas's tests for being a rational discourse agent, that of self-awareness or being human.

Whether this recognises those who cannot speak – those human beings in states of noncommunication – or those who choose to communicate purely through actions, whether of compassion or violence (the silent monk or the silent bomber), is moot and perhaps indicates a flaw in the theory; Habermas sees being human as something to be worked at together. Those who kill – and Habermas defines terrorism, for example, as an inexcusable act (i.e. one without legitimate reasons that might 'excuse' it) – are, by refusing to enter into discourse, in fact refusing to admit their humanity (Borradori 2003). Habermas, although he does not state this as such, seems to regard political (and presumably religious-political) acts of murder, or terrorism, as being without that human motivation that, while unable to excuse the act, makes it capable of seeming 'human', in that it arises out of motives that can be understood. This is perhaps too broad a point, for surely even fanatical reasons can be understood. It may point to an idealistic aspect of the species ethic, specifically that we should choose to enter into moral discourse over simply deciding that the enemy is only capable of comprehending force or should be killed rather than discoursed with.

What Habermas might be saying is that to argue for religious or sociopolitical reasons as justifications for murder may fail the validity test, in other words, may be illogical (the obvious illogicality being that killing will be equal to salvation) or irrational. Habermas might have looked to Dostoevsky for an interesting illustration of this point. Can one murder for rational reasons? An entire Russian historical set of assassins predicated this very point; and Dostoevsky's Raskol'nikov in *Crime and Punishment* (1866) incarnated the issue of whether one might murder for pragmatic and rational reasons based on the redistribution of wealth. To Raskol'nikov as a 'modern man' free from emotions and superstitions, capable of acting with calm pragmatism to redress what he perceives as social injustice, it is something of a shock to discover that his conscience will not allow such a 'validation' of his act, and the sheer horror of the assault overwhelms him. His failure to behave rationally is the point of Dostoevsky's moral debate.

Habermas is perhaps a little idealistic in arguing that the most human quality is that of wishing to enter dialogue, given that different political and religious beliefs may lead one to refuse this condition. And if you do not wish to do so, implying that you are not fully 'human', this contravenes Habermas's rule that discourse should be inclusive.

The basic impulse in the human being must be to submit his/her thoughts and beliefs to reasoned validation; yet what of those who will not even admit the usefulness of rationally examining their own as well as others' beliefs?

We now move from the comforting universal context where 'we are all human and thus capable of discourse' to the more problematic area in which relative types of awareness and knowledge may or may not prioritise reason. Some cultures place greater stress on intuitive or belief-based forms of knowledge (the primacy of religion in such countries perhaps offering a sharp indication of their cultural norms).

The desire to communicate may be the truly universal human trait; however, it implies that virtues may be more relative.

6.2   Reason as a Universal Virtue?

What does reason know? Reason only knows what it has succeeded in learning (some things, perhaps, it will never learn)…you tell me again that an enlightened and developed man, such, in short, as the future man will be, cannot consciously desire anything disadvantageous to himself…. But I repeat for the hundredth time, there is one case, one only, when man may consciously, purposely, desire what is injurious to himself, what is stupid, very stupid – simply in order to have the right to desire for himself even what is very stupid and

> not to be bound by an obligation to desire only what is sensible...choice can, of course, if
> it chooses, be in agreement with reason.... But very often, and even most often, choice is
> utterly and stubbornly opposed to reason.... (Dostovesky 1862)

Every subject with the competence to speak and act is allowed to take part in a
discourse. What is a 'competent speaker'? Such a person must be:

1. Voiced (able to speak)
2. Be free to say yes or no
3. Able to open all statements to questioning, modification, and confirmation as
   well as to agree which interests make legitimate claims on all and which do not
   (able to overcome egoism)
4. Capable of mutual understanding
5. Most importantly, committed to rationality in discourse (Rehg 1994)

Rationality is a tricky notion, given that it implies emotional neutrality – not
always a prevalent human trait. Rawls's famous idea of an objective position 'from
behind the veil', one without bias, for making decisions based on the social and
economic good, arguably fails for two reasons. Leaving aside the question of
whether the original position could be replicated, the assumption that one might
automatically choose a society that optimises aggregation, that people would
behave in a risk-averse way (i.e. 'concerned to make the worst-off position as good
as possible') to maximise the average position, is not proven (Swift 2006). We are
not necessarily going to adopt as our good the *rational* view of what is good for
the whole of society.

At the core of Habermas's model is the idea that the agent will be able to act in a
rational way during dialogue, capable of tempering his/her own prejudices and
achieving consensus. Most readers will remember instances when they and others'
discussions have been proven irrational, a contest for sheer bloody-minded domination, compromise no longer on the agenda. Habermas's view of human nature, and
of the moral development of the human being, may expose a flaw in his notion of
the probability of virtuous discourse. However, when Habermas states that motives
for consensus are complex in derivation, he points to this potential grey area in his
reasoning, intimating that a norm, or consensus on behaviour, is based on reasoning
but also on 'force':

> Normative claims to validity, then, mediate a mutual dependence of language and the social
> world ... gaining acceptance on the part of a norm is encoded in a twofold fashion because
> our motives for recognising normative claims to validity are rooted both in convictions and
> in sanctions, that is, they derive from a complex mixture of rational insight and force.
> (Habermas 1991a, b)

Is a discourse's agent's rational 'virtue' different from individual motives that
may be driven by need or desire? This would seem a rather idealistic view of human
nature.

When he discusses the power of rational language to establish a morally significant connection, Habermas (1991a, b) not only discusses the reasons that might be
given for a statement to make others believe it but also suggests that we must grasp

the *motives* behind the statement and the soundness of those motives. The distinction between reason and motive suggests that there is a foundation for reason, as indeed we see from the species ethic that motivates discourse agents both to understand the other in as rational a way as they can while explaining themselves rationally. The ethic is in fact an innate (species) virtue that motivates the desire and the need to engage in dialogue (Glenna 2008). Thus, Habermas's species ethic may be expressed in various ways as virtues or motivations behind the action, the desire, or the need to act in a certain way, for example, as altruism, the wish to promote autonomy, or even as love. It may be expressed differently, such as when 'love' may be a broad notion, one influenced by social, cultural, and historical contexts. A Habermasian virtue is therefore a broad idea, as we can readily appreciate from his dislike of the term. Although everything must be subject to reason, the sources of that 'everything' are not necessarily rational. Certainly one's knowledge is often highly influenced by emotion or intuition. Thus, according to Habermas, 'reason' or 'rationality' can be understood in three different ways:

(a) As a procedural notion (i.e. 'reasoning') applying to the practice of dialogue or discourse and requiring interlocutors to provide what evidence they can to argue their views and beliefs
(b) As a trait or virtue, implying the interlocutor's desire to achieve compromise by submitting his/her own values and ideas to scrutiny within reasoned discourse
(c) As a culturally charged context for the interlocutor's upbringing and personality, meaning that reason implies an attitude towards the *value* of rationality – and that this may become a factor that impedes dialogue

Point (a) is not up for debate in Habermas's view, since discourse requires scrutiny through reasoned debate. However, (a) and (b) relate to how knowledge enters into discourse and to how one might cultivate reason as a potentially universal value.

Dostoevsky, that great student of human nature, had a very 'Russian' view of human nature as irrational, both positively and negatively so. Whereas the interpolation of the Russians may seem strange at this point, it is not particularly so given the eternal debate among Russian scholars as to whether their country belongs to the West or the East – that is, whether it is a product of Catherine the Great's enlightenment (given her admiration for all things Diderot) or if it remains stubbornly 'Eastern' (characterised as chaotic, emotional, and prone to self-abnegation). To Dostoevsky, perversity is the salient trait of the human being. The notion of a utopia being ultimately boring to man, so that he will start to 'stick pins into himself' (as he suggests in his 1862 manifesto to the perverse, the novella *Notes from the Underground*), does not bode well for rational attempts to construct such a utopia. Dostoevsky tests the limits of human nature; he liked his extremes and chose themes through which his protagonists might be forced into extreme behaviour. His point was that human beings often tend to act in surprising, not rational, ways.

Against such an argument, Habermas's views seem rather dull and even improbable. However, the first point to note when discussing Habermas's views of rationality is that communicative action is tied to reason, which he sees as a capacity inherent within language, especially in the form of argumentation. The structures of argumentative speech – which Habermas identifies as the absence of coercive force, the mutual search for understanding, and the compelling power of the better argument – are what intersubjective rationality uses to make communication possible. Actions undertaken by participants in a process of such argumentative communication can be assessed as to their rationality by the extent to which they fulfil those criteria. In short, one does not have to be *entirely* rational to speak rationally. The need to persuade forces one's emotions into logical structures. We speak not to rant to ourselves, but to be heard; speech that does not imply an interlocutor is the speech of the mad.[4]

Speech, as Habermas states, is indeed the social and rational means by which we live together.

We may also end up with a form of passionate argument, as per Mackie's (1976) view that impartiality is not as effective as a more invested approach to discourse. As long as what we say is subject to reasoned scrutiny, any emotion behind it may make us try harder to make what we say *more* rational, rather than less. Annas (1993) posits that virtue involves a deeply personal commitment to that point of meaning at which one's life is aimed, that ethical beliefs have to become rooted in one's emotional life before they can be effective.

Yet belief is not the same as rational argument, and while the claim that knowledge is belief that is true and justified sounds attractively simple, the alleged lack of recognised methods for settling moral disputes suggests that belief cannot be easily 'justified' (Attfield 1995). And as Nussbaum (1992) has noted, evaluation of facts can even be a matter of power, self-assertion, or utility.

Solomon (1995) argues that we need to care to achieve justice; 'justice is not an ideal state or theory but a matter of personal sensibility, a set of emotions that engage us with the world and make us care – as reason alone … cannot' (p. 197). Solomon's (1993, 1999) 'rational Romanticism', his view that emotions are judgements rather than blind or irrational forces, offers a way forward on this issue. He argues that rationality is the product not only of thought but 'also of caring', that an emotion 'is a system of judgements, through which we constitute ourselves and our world', for

> …emotions constitute the framework (or frameworks) of rationality itself… together our emotions dictate the context, the character, the culture in which some values take priority, serve as ultimate ends, provide the criteria for rationality and reasonable behaviour. Our sense of justice… [is] a systemic totality of emotions, appropriate to our culture and our character, that determines not only particular emotions…but also the standards and expectations according to which those emotions are provoked. (Solomon 2003)

Thus, compassion, pity, and sympathy are not just 'feelings', 'they are also engagements in the world, instances of involving if not identifying oneself with the circumstances and sufferings of other beings', for 'compassion and its kindred emotions focus our attention on the world, on the person or creature who is suffering' (Solomon 1995).

---

[4] 'Talking to oneself' arguably still implies an interlocutor….

Habermas tackles the issue of the invested participant by reminding us of human fallibility:

1. He suggests that emotion *does* have a part to play in the participant's involvement.
2. He puts forward a different way of looking at the participant's self-view or identity.

On (1) above, Slote argued in the 1980s for 'dependent' virtues, those attaining full status as desirable traits when 'accompanied by' (i.e. dependent on) other 'desirable traits' (Slote 1983). What we are looking for, it seems, is a harmonious collection of interdependent virtues, one that moreover acknowledges limitation or imperfection. To this end, Christine Swanton (2007) offers a view that a virtue is a 'disposition to respond to or acknowledge…in an excellent or good enough way'. 'Good enough' is a usefully practical amendment of the idealised notion of the altruist (and conforms to Aristotle's claim that the subject matter of ethics does not allow for a high degree of exactness, that we 'must be content to draw conclusions that are true only for the most part') (Hursthouse 2011).

Habermas's critics have pointed out that his theories are too abstract and so do not altogether explain how one might produce valid moral judgements, to which Habermas (1993) has replied by noting the element of fallibility in judgements, so that they can only ever be provisional.

Another way of looking at the notion of fallibility involves noting the shift away from the theory of the Ideal Observer (an impartial, omniscient, ideal judge of morality), one who unfortunately presents us with 'unattainable, unknowable standards' (Kawall 2006). The virtuous discourse agent may even be a form of Ideal Observer, a 'knowledgeable, impartial, and consistent person' (Brandt 1995).

Korsgaard (2008) argues that emotions are important to the Ideal Observer. She interprets Aristotle as suggesting that emotions contribute to rational activity, that emotions are 'a kind of *perception* of the good'.

The Ideal Observer is less neutral and rational than usually thought. Swanton (2007) argues for a replacement of 'virtue agent as oracle' (the Ideal Observer approach) with a 'virtue-ethical species of dialogic ethics', for 'the aim of an agent in attempting to solve a problem is to integrate the various constraints on its solution' with the aim of maintaining an overall 'rightness' to the (modified) solution. The Ideal Observer is a theoretical device, not a practical reality; however (Kawall 2006), it is often replaced by the wise 'expert', both knowledgeable and skilled at gaining information in situations 'where human interests and perceptions are paramount' (Walker 2003). This expert is endowed with 'appropriate emotional sensibilities and context-sensitive practical wisdom' (Swanton 2007). The shift from impartiality to 'emotional sensibility' indicates a movement towards a relational process (instead of a detached and hierarchical one), through which mutual understanding is achieved. Rather than a discourse process with one 'ideal' expert, the discourse process invites all equally to work towards an ideal solution, which is likely however to be tempered by consensus. Rather than 'thinking for others', what is advocated is a process of 'thinking *with* others' (Walker 2003). We do not need a wise agent, but wise *agents*, whose view of self within the discourse process is both dualistic and more fluid – a concept discussed further on.

In his *Republic*, Plato (2008) lists the four cardinal virtues as wisdom, justice, courage or fortitude, and moderation or temperance. For Aristotle, temperance was the ability to find the mean between a state of excess and one of deficiency, analogous to the more modern idea of rational neutrality. Temperance, or neutral impartiality, indicates a continuation of virtue across the centuries, though it is perhaps interpreted differently. Does Habermas, like Aristotle, believe that the virtuous man should possess a certain 'courage and persistence' in working towards that good, 'to overcome the numerous obstacles to an adequate and shared understanding' (Swanton 2003)? In other words, does he argue his point presumably with consistence as well as persistence?

Discourse agents must demonstrate open-mindedness, the ability to comprehend other points of view as well as complex information, in line with Habermas's/argument that interlocutors should be open to the validity claims of all who are engaged in the dialogue and may change their beliefs on the strength of evidence and publicly acknowledged mistakes – thus they must have the capacity to learn.

Swanton (2003) allows for virtues that are dynamic (i.e. 'can be improved') and suggests 'a dialogical method for constructing solutions … so problems can be identified and addressed' in a way which allows for a variety of voices or perspectives. She argues that Habermas puts forward 'an ideal which enables us to evaluate critically the institutional and interpersonal conditions under which what is right is currently determined, not a specification of how such judgments should be reached in concrete circumstances'. Siep (2005) sees the process of virtue development as fluid, a process of continual renewal and discovery of admirable behaviour. This is not necessarily conflated with Aristotle's belief in consistency, a stable state, or 'appropriate rigidity' of character, although it should be noted that Aristotle's sense of virtue does imply a developing state, for the virtuous person learns from past choices (Sherman 1999). However, Aristotle's flexibility is linear; Habermas's is part of the ongoing flux of pluralistic dialogue. For Habermas, a courageous flexibility is a defining characteristic of the discourse agent, along with his/her capacity for practical wisdom.

A similar pragmatism informs the notion of consensus itself. If the 'truth condition of propositions is the potential assent of all others', this implies a universal audience – but what is the knowledge level of this universal audience meant to be (Habermas 2001)? No universal audience can, for example, have universal in-depth scientific knowledge.

It would seem that we must return to a nonideal situation, in terms of which a nonuniversal, but sufficiently inclusive, group works together following a set of validity rules that test their statements for rationality and acceptability. Situational knowledge becomes important to the making of a valid judgement (Blaug 2000).

The shift from Ideal Observer to wise expert with his/her 'appropriate emotional sensibilities' also raises the issue of emotion as a necessary complement to reason.

One can see why philosophers distrust emotions, due to issues of bias, temper, and unwillingness to engage in debate. Stansbury (2009) quotes the example of a local campaign against a hazardous-waste incinerator, in which local activists refused to engage with the scientific or economic reasoning behind the project and instead embarked on emotional letter writing, vilification, and even violence – all counterproductive activities.

Yet a lack of emotion may be just as bad as too much.

Aristotle's golden mean, or temperance principle, surely implies not 'no emotion', but the 'right amount of emotion'. Broadie (1991), discussing Aristotle's division of virtues into the groups of character and intellect, suggests that Aristotle viewed virtue as not merely the result of knowledge but also of appropriate *emotional* responses. Solomon (1993) makes much the same point, arguing that rationality is about having the 'right emotions', 'caring about the right things'; appropriateness is important because 'to say that emotions are rational is to say they serve purposes'.

Emotion is modified by reason, becoming a useful mean between extreme involvement and dispassion. To Aristotle, one's feelings should be in a correct harmonious ratio to one's virtuous judgements. Of course, we must always avoid merely chaotic emotional states; however, the fact remains that for Aristotle a virtuous person does not act from an absence of feeling (Annas 1993). There is some modern agreement, such as in Slote's (1983) argument that emotional goals are to be considered alongside the rational. Burnor and Raley (2011) suggest that virtue ethics allows for a reconciliation of impartiality and personal feeling, while Lovat and Gray (2008) have argued that virtue ethics offers a middle way between deliberative approaches and those based on intuition.[5]

Precisely what emotions – or types of knowledge or cognition – are we talking about then if 'intuition' is favoured by some philosophers? And what does Habermas have to say on the subject? This will be dealt with in Sect. 6.3, but the issue of emotion is still pertinent.

In this context, it is more useful to look at a modern Habermasian commentator such as Seyla Benhabib, who, as Hutchings (2010) neatly summarises it, sees discourse ethics as requiring more 'sensitivity':

> Benhabib argues that discourse ethics is not a way of 'cutting like a knife' between claims that can be accorded universal validity and those that cannot, but is rather a form of moral judgment that combines respect for a principle of universalizability with the capacity to recognize and be sensitive to difference.

Benhabib suggests that rational scrutiny should be balanced or complemented by sensitivity to difference. Her version of discourse ethics implies an attitude to discourse that links with the species ethic and reason to form a neat trio: the willingness to speak, to discuss rationally, and to accept with good will the range of differences that are likely to be subject to reasoned debate.

Iser (2003) argues that discourse requires 'sensitive perception'. By 'sensitivity', he appears to mean a willingness to assist and thus an openness to dialogic viewpoints, or what he calls 'good will', while Hursthouse's (1999) appeal for a virtue

---

[5] See also Lovat (2003).

ethics of hope arguably posits the need for emotional optimism. She also argues for empathy, stating that 'our understanding of what will hurt, offend, damage, undermine, distress or reassure, help, succour, support or please our fellow human beings is at least as much emotional as it is theoretical'.

The feminist ethics of care developed by Gilligan and others contends among other things that notions of impartiality and universalisability, or abstracts such as justice, can denigrate the more particularistic attachments between individuals. The ethics of care recognises situational complexity over more general ethical rules applied to general, not specific, situations. According to the ethics of care, it is morally counterintuitive to ignore individual differences in favour of an impersonal right. The justice perspective and the care perspective are two different ways of organising one's moral thinking. Challenging the Kantian view that moral action is motivated by respect for universal laws, the ethics of care makes room for actions motivated by compassion and sympathy – the altruistic emotions as one might call them.

Benhabib's expansion of Habermasian discourse ethics borrows from care ethics. Arguing against Habermas's restriction of moral concerns to universalisable questions of what is right for all, she argues that he conflates the 'standpoint of a universalist morality with a narrow definition of the moral domain as being centered round "issues of justice"'. The context of deliberations may, in fact, require consideration of the particular needs of the other, established on the basis of care. Her view is that Habermasian impartiality and universalisability claims deny difference. To be morally valid is to be equally good for all. For Benhabib, to be ethical is to adjust one's understanding of whether a morally valid norm is in fact equally good in that specific instance where there may be negative consequences for a concrete other. In other words, principles should not be applied to all people indiscriminately.

Habermas recognises that care for the other is a necessary condition of discourse, yet the claims of care appear to be subsidiary to universal justice in his system. Since universal justice is a problematic term when looking at global virtues that might apply to global or universal discourse, another way of framing it might be to say that Habermas notes how in discourse, a norm can only be justified with reference to a reason external to the individual – thus subjectivity is balanced by external reasons (Keller et al. 2005). The other in discourse accepts the validity of a statement because they accept the good reasons for that statement (a validity claim has a 'warranty' as it were, assumed behind it). However, a persuasive argument for a norm must tie the norm to a language of wants and needs (Rehg 1994). Interlocutors in discourse might be swayed by emotional persuasion more than rational argument – but Habermas's riposte would be that, ultimately, there would have to be a *logical* discussion of the norm and its consequences.

In discussing the issues of subjectivism and relativism in applied ethics, Tim Dare (2002) argues for 'expertise in ethical reasoning', achieved through proficient reasoning skills, knowledge, and a commitment to understanding and finding reasoned solutions – one should be skilled in constructing and assessing reasoned support for any ethical position. What makes an ethical judgement *correct,* according to Korsgaard (1986), is that *endorsing* that judgement is constitutive of rational, reflective agency.[6]

Habermas (1991a, b) argues that the appropriate application of moral norms to specific situations requires a sense of empathy (*agape*). He later defines this as a form of 'considerateness', meaning awareness of intersubjective feelings. This 'empathy', as Habermas might define it, is a distinctly moral form of perception; it is essential to the process of dialogue.

Thus Habermas's view of reasoning is balanced by the understanding that humans have feelings, 'wants, and needs' that inform the process of reasoning. It can be argued, as Brentano (1973) has done, that an emphasis on feeling, without lapsing into subjectivism, can be conflated with the idea that there is a 'higher class of feelings' or 'emotional activities' common to all, identified with a form of love that places value outside of the self onto an object or activity.[7] This has some similarity with the Habermasian view of transcendence, as will be discussed later.

The conclusion so far is that emotion is a pre-context to the process of reasoned debate and one that may in fact encourage the species ethic in us – the desire to enter into communication and the need to develop 'considerateness'. Emotion adds harmony, care, the passion behind reasoning, or Solomon's 'judgement' capability.

What of Eastern views on reason? Eastern rationalism seems always to be a subset of harmonious unity, mind balanced by heart. Confucianism, central to the development of Chinese thought, supports belief in the notion of order or harmony (*li*):

> There is a prevalent generalization that takes the Western culture as rational and the Chinese culture as emotional. In fact, what the Confucian school emphasizes is the general 'li' as reason that deviates from the concrete 'qing' as emotions... The ideal realm the Confucian school pursues is a state of 'rationality', 'harmony between reason and emotion', and 'harmony' between heavenly principles and human feelings. (Ch'ung and Zhen 2003)

Perhaps this is close to the idea of sincerity, in which mind and heart might be in balance. Habermas's view of discourse admits this term, appearing to use it in the sense of a tool for gauging falsehood in discourse, but nonetheless, sincerity suggests an emotional parity between the discourse subject and how one feels towards it. Admitting, for example, that validity claims are often made on the basis of subjectivity (feelings, moods, desires, beliefs, and the like), he argues that such claims are:

> ...open to rational assessment, not in discourse but by comparison with the actor's behavior: for example, if a son claims to care deeply about his parents but never pays them any attention, we would have grounds for doubting the sincerity of his claim. Note that such insincerity might involve self-deception rather than deliberative lying. (Bohman and Rehg 2014)

---

[6] See also Korsgaard (1996).

[7] See also Donohoe (2004).

Habermas's view of emotion therefore relates to procedural issues such as coming to a dialogue with a sincere and committed mindset that allows for empathy or 'considerateness'.

Asian countries do not all follow one particular set of virtues, though with the possible exception of Japan,[8] most fall broadly into the Confucian area.

It can be argued that Confucianism has little in common with Western virtue ethics. Ames and Rosemont (1999) distinguish a Confucian role ethics from Western virtue ethics for, like Nuyen (2006), they see the community as prioritised over the individual. Yet the idea of individual and community virtues being compatible is surely not too far-fetched an assertion, if we examine the interaction between what can be called the more individual Confucian virtues and the more 'socially' oriented ones.

Van Norden (2007) argues that Confucius may not have had a list of cardinal virtues, but his follower Mengzi did, such as benevolence (*ren*), righteousness (yi), wisdom (*zhi*), and propriety (*li*). Other critics suggest two forms of Confucian ethical clusters, one based around notions such as respect for parents, loyalty to government, and keeping to one's place in society; the other being the 'central Confucian doctrine' of *ren*, 'humanness', or 'care'. This might seem somewhat analogous to Habermas's view of humans as being sufficiently caring to engage in discourse.

As Van Norden (2007) relates, *ren* implies both dutifulness (*zhong*) and reciprocity (*shu*). *Ren* is a specifically 'relational' term, meaning an implied attitude to others (Hui 2004).

Ren, sometimes translated as love or kindness, is not any one virtue, but the source of all virtues. The Chinese character literally represents the relationship between 'two persons' or 'co-humanity' – the potential to live together humanely rather than scrapping like birds or beasts.

Ren keeps ritual forms from becoming hollow; a ritual performed with ren has not only form but ethical content; it nurtures the inner character of the person and furthers his/her ethical maturation. Thus, if the 'outer' side of Confucianism is conformity and acceptance of social roles, the 'inner' side is the cultivation of conscience and character (Li 1999).

Other commentators have added *chung* (loyalty to one's true nature (Wenstop 2005), but also meaning exerting one's best efforts to serve others and, as such, forming a subset of reciprocity, or *shu*) (Yee 1999).

Ren ethics seems fairly similar to the Habermasian species ethic on the following points:

1. It reveals itself in a relational form.
2. It requires sincerity or 'inner form'.

---

[8] The Confucian influence on Japan is, of course, significant, but in terms of current thinking, perhaps less so: 'In other countries, including modern China, contemporary interpreters are examining Confucianism as a living philosophy of ongoing significance, but Japanese scholars have more viewed it as a historical artifact, not a vital philosophy'. See Tucker (2013).

3. It is linked to ethical maturation – a concept Habermas picks up in his view of human development and how the human being progresses through various stages of knowledge.

The ethics of care, *ren* ethics, or the Western notion of charity, all imply a human mutuality capable of universal acknowledgement. To illustrate this, we might use a term from African ethics, that of 'ubuntu', a major tenet of African ethics.

6.2.1   Ubuntu[9]

Ubuntu has been suggested as uniquely universal, for:

> it emphasizes respect for the non-material order that exists in us and among us; it fosters man's respect for himself, for others, and for the environment; it has spirituality; it has remained non-racial; it accommodates other cultures and it is the invisible force uniting Africans worldwide. (Enslin and Horstheke 2004)

Eze (2010) notes that the idea is less of an imposition of a culturally homogenous norm than an allowance for contradictions and differing contexts, as to 'understand ubuntu... is therefore to locate the context in which it was invoked and recognized as a normative rule governing social practices', for there can be no homogeneity, as social norms differ in differing contexts or communities. Thus, *ubuntu* is interculturality and a new humanism that moves beyond colonial power relationships to the simple humanity of mutual creation – 'we create each other'.

One is human because one belongs, participates, and shares. In *ubuntu*, a central concern is the healing of breaches, the redressing of imbalances, and the restoration of broken relationships, all while seeking to rehabilitate both the victim and the perpetrator.

Gichure (2006) suggests that ubuntu combines personal responsibility and common good, for as the African proverb has it, 'a person becomes virtuous through the virtue of others'. This comment opens up the central tenet underpinning much of Asian (bio)ethics: that of the mutual recognition of each other's 'humanness'.

It has been argued that African ethics is an amalgam of Western and traditional values, that 'ubuntu', 'the foundation and the edifice of African philosophy' (Ramose 1999), 'the basis of African communal cultural life' (Tambulasi and Kayuni 2005), 'functions as a unifying factor, bringing people together regardless of their background or status (Sithole 2001). It teaches unity – 'above all, that we are a collective with the success of one person depending very much on the success of all, in terms of combining personal responsibility and common good' (Gichure 2006). *Ubuntu* implies a recognition of one's own humanity in the other (Luhabe 2002) for 'a person is a person through other persons' (*umuntu ngumuntu ngabantu*) (Shutte 1993). Bishop Tutu (1999) explains it thus:

---

[9] See Wiredu (1996) and Broodryk (2007). For a negative view of ubuntu, see Kresse (2011).

When we want to give high praise to someone we say, 'Yu, u nobuntu'; 'Hey, he or she has *ubuntu.*' This means they are generous, hospitable, friendly, caring, and compassionate. They share what they have. It also means my humanity is caught up, is inextricably bound up, in theirs. We belong in a bundle of life.[10]

The above suggests the nexus for Eastern *ren* and Western individualism: 'I' requires 'you'. Or as Habermas might put it, to be human means having someone with whom to discourse because our species ethic requires that communication.

What else does discourse require in terms of virtue?

6.3 Emancipatory Knowledge

Human beings, as a species, are capable of harmony, both within themselves, and with each other. (Hursthouse 1999)

Let us take a step back and begin with one aspect of virtue that engenders little debate: practicality.

Broadly speaking, there is one obvious similarity between the theories of discourse and virtue ethics; both Aristotle and Habermas place the burden of truth on individuals making rational judgements. Habermas's *praxis* is arguably a version of Aristotelian *phronesis,* or practical reasoning needed for ethical action. To Aristotle, a virtuous person is above all distinguished by practical wisdom – this implies *both* character and action. Virtue ethics today is less distant from deontological and consequentialist theories than might be thought, in that although the focus remains on the virtue of the agent, the actions of the agents can be evaluated in terms of their contribution to the preponderance of the good or to consensus (Sandler 2007). Likewise, Hursthouse (1999) has reminded her readers that virtue ethics is also act centred, not merely agent centred, and one might also note Ludwig Siep's (2005) argument that virtue can lead to intersubjective convergence, in that the agent and the receiver confirm the virtue of the act. Opinions differ not only on what constitutes a virtue but on what constitutes a virtuous act: Rosalind Hursthouse considers a virtuous act to be what a virtuous agent would do; Michael Slote has narrowed that further, suggesting that a virtuous act is what a virtuously *motivated* person would do, while Christine Swanton argues that a virtuous act is one that realises the end of virtue. But, in short, the virtuous person is required to demonstrate, not merely possess virtue.

Aristotelian virtue is 'a state of character concerned with choice' (Aristotle 1998). To Aristotle, a virtuous person takes pleasure in choosing to do the right thing, as well as in doing it temperately and habitually, so that 'we become just by doing just acts, temperate by doing temperate acts, brave by doing brave acts' (Aristotle 1998). Wisdom must therefore be demonstrable through practical acts.

---

[10] See also Lutz (2009).

To a degree, this mitigates the complaint made about wisdom being somehow related to elitism (i.e. through education). Aristotle's elitist view of those with education and wise virtue making decisions for others is clearly noninclusive – a major difference compared to Habermas, for whom inclusivity is essential in dialogue. However, a modern interpretation of 'wisdom' might view it as being culturally determined; an example would be a dialogue in which a Harvard or Oxbridge-educated interlocutor is grouped with an African tribesman. One might argue from the standpoint of Western wisdom, the other from generations of tribal folklore and proverbial knowledge. There is not necessarily a hierarchy of wisdom, merely a different *context* to each.

According to Habermas (1979), communication works because there is an internal relation between meaning and validity. He suggests that in making any utterances, the speaker raises four validity claims, on which he/she can be challenged:

1. The speaker can be challenged on the meaningfulness of what he/she says.
2. On the truth of the facts about the world he/she assumes.
3. On his/her authority to make an assertion.
4. On his/her sincerity.

Point (4) has already been accepted in terms of Habermas's tendency towards accepting the interlocutor's investment in the discourse. A sincere interlocutor is a basic element in his notion of admitting one's species ethic driven desire to enter into dialogue.

In terms of (1), it may be argued that questions about the nature of meaningfulness add to the challenge. However, one might stick to a practical definition of meaningfulness as that which can be reached through reasoned debate, that is to say, it is factual and thus agreed upon by interlocutors, making it a communal decision.

Regarding point (2), we must ask what happens when the facts are open to differing interpretations, as can happen with new and untested sciences such as nanotechnology.

Even point (3) about authority is problematic if it is assumed that the speaker is self-deceptive about his/her right to make a validity claim. The notion of a speaker who is always sufficiently self-reflexive to admit his/her deceptive authority is rather an idealistic one. (Or, in short, if we knew when we were being stupidly pigheaded, we'd stop and agree that others might have a point – something that anyone who has ever got into a trivial shouting match knows to be extremely hard.)

Yet Habermas suggests that dialogue will expose the speaker's right to say he/she is right, and as such embeds the notion of nonhierarchical, pluralistic, and fluid debate into discourse. Any assertion, he believes, can be checked within such a structure. The degree of self-deception is perhaps the greatest issue in terms of compromise, or the ability to accept one's 'checked' assumptions.

Self-deception is a key notion for Habermas. He differentiates three cognitive areas in which knowledge is generated:

- Work knowledge, or the way one controls and manipulates one's environment.
- Practical or social knowledge, governed by binding consensual norms, the validity of which is grounded in intersubjectivity, i.e. the mutual understanding of intentions.

- Emancipatory knowledge, 'self-knowledge', or self-reflection. One becomes the critically reflective knower who knows the self as the person doing the knowing (Gray and Lovat 2007).[11] Insights gained through critical self-awareness are emancipatory in the sense that at least one can recognise the correct reasons for his/her problems. Knowledge is gained by self-emancipation through reflection, which leads to a transformation of one's perspectives.

Thus, one might note that there is now another way of looking at sincerity – namely, to see it as the antithesis to self-deception and as a major aspect of emancipatory knowledge. Accepted as such, one's sincere and humbling acceptance (e.g. of one's own mortality) might be a universal one, as Veatch (2004) suggests:

> A recognition of human finitude leaves one simultaneously affirming a single universal moral authority and a deep sense of human inability to know the content of that moral authority in a definite way. The impact on cross-cultural ethics is critical. We can simultaneously affirm a common morality and show respect of the differences in those cultures that do not share our own moral perception. Finitude requires respect for the moral views of others without surrendering one's conviction that there is a single, universal foundation for morality.

The emancipated, self-aware self is humble, open to reason, and willing to compromise. He/she is also capable of *dialectical agency*.

6.4　Dialectical Agency

The dialectical process of (in simple terms) arriving at the truth by stating a thesis, developing a contradictory antithesis, and before combining and resolving them into a coherent synthesis suggests that the dialectical process and dialogue have much in common. The term 'dialectical process' is often used widely as any two oppositions that must be synthesised for progress; a dialectical process can mean a creative tension between opposites.

The real East-West difference may be in terms of the dialectical process implied by the Habermasian model, one in which self and other exist in a dialectical state. The Confucian model implies that such a state must evolve towards the goal of social harmony; the Habermasian, towards the goal of consensus. There appears to be a *difference in terms of how much dialecticism is allowed,* in that Western individualism encourages greater opposition to the whole. In short, through dialectical process, agreement would be partial, a compromise between agreement and disagreement. This implies that the discourse agent, to achieve emancipatory knowledge, must exist within the dialectical tension between relative and universal values, between the goal of the common good, and that of individual and contextual desires.

---

[11] See also Mendieta (2002) for the argument that Apel's philosophy has a focus on self-reflexivity.

There are several forms of dialecticism within discourse. For example, there is an essential distinction between the motives of the interlocutors (participants in dialogue/discourse) and the procedural aims of the discourse or dialogue. This implies that the interlocutor performs a dialectical role, between his/her own motives, needs, and virtues and the procedural requirements of the discourse, adapting his/her views as the procedural imperative requires. There is a dialectic between relative/subjective views and the normative aim that such views might impede, with a common criticism of virtue ethics being that it typically appeals to emotive and culturally influenced notions such as justice and courage and so does not 'provide a standard that successfully distinguishes virtue from vice' (Conley 1988). In other words, it provides only a form of relative ethics that is insufficiently useful in terms of binding a group.

In Habermas's model, relative and universal ethics exist in *a necessary dialectic*. Dialecticism is in fact the crucial factor, rather than 'wisdom' or knowledge, or indeed any issue concerning educated elitism.

The notion of relational dialectics in communication theory, developed by Baxter (1988) and Rawlins (1988) in the 1980s, suggests relational communication as a clash of desires. For dialogue to be useful, it must in fact be dialectical, i.e. opposed, multivoiced, and thus capable of evolution.

6.4.1 Inclusive Dialecticism and Dialogue

> Truth is not born, nor is it to be found inside the head of an individual person, it is born between people collectively searching for truth, in the process of their dialectic interaction. (Bakhtin 1999)

There is a definition of dialogue derived from the work of Russian critic Mikhail Bakhtin that offers further nuance to the idea. In Bakhtin's (1981) view, any utterance is formed through a speaker's relation to otherness (other people, others' words and expressions, and their lived cultural world in time and place). When he stated that the word 'lives, as it were, on the boundary between its own context and another, alien, context', Bakhtin (1986) implied an inherently dialogic quality in any speech act that segues into a dialectic spiral of understanding/compromise and then higher understanding:

> Otherness means that any word uttered requires an answer: Any understanding of live speech, a live utterance, is inherently responsive… Any utterance is a link in the chain of communication.

Every statement contains in fact the echo of its own opposition, since speech is inherently oppositional. Thus, in discourse, one says what is true to oneself, but also predicates the echo of an opposing view, the voice of the other. As Bakhtin (1981) stated, 'the word…exists in other people's mouths, in other peoples' concrete contexts, serving other peoples' intentions; it is from there that we must take the word and make it our own'.

In order to achieve a state of agency in which one can act as a global citizen, one needs to achieve an intersubjective state through dialogue that implies the validation of one's words only through the speech of the other. Identity is reliant on otherness, for 'a living person is able to do furious battle with definitions of their personality in the mouths of other people' (Bakhtin 1984).

The hardest part of the dialogue process is the recognition of the validity of others' viewpoints, namely, the undermining of the belief in the rightness of one's own views. A successful dialogue requires a dialectical shaking of the 'certainties' of one's identity; this is even more difficult when undertaken globally. Such a process is reliant on the goodwill of the agent and, most significantly, his or her ability to accept the pluralism of truth and of the self, his 'dialectical agency'.

Refining a dialectical process in the Bakhtinian sense may prove to be a way forward for improving global dialogue on new technologies such as nano. The fundamental principle of a Habermasian pTA forum intended at global consensus operates on a simple (yet complex) procedure, that of requiring all agents to argue the opposing position(s). From such a dialectical process, it is hoped that consensus may be reached.

The virtuous agent will search for good through dialogue, balancing reason and empathy, as well as private and public selves. Aristotle's concern with balancing the demand for absolutes, with the need to trust our sense perceptions, suggests an innate dialecticism in any virtuous dialogue leading to decision-making.

Thus, Habermas brings discourse ethics down to the individual virtuous discourse agent, whose chief virtue, it seems, is that she or he is self-aware. But this suggests the familiar virtue debate between Western individualism and Eastern community-mindedness. It appears we have come full circle to the original problem without making any particular headway. Or have we? Perhaps the conclusion so far is not a question of emotive reasoning, or of what one knows, but of *how* one knows it – whether one is able to admit of a dialectical model of self-awareness or not.

Or to express it rather simplistically, in the West, the individual's knowledge enters into full dialogue with society. By contrast, in the East, since dialogue is entered into with the full knowledge that society must ultimately be seen as right, harmony must be restored, and subjects are less free to admit of dissent, dialectical agency is only partially achieved.

We may be back at the question of autonomy.

6.5 Autonomy as a (Non)Universal Virtue?

> The end of man is…the highest and most harmonious development of his powers to a complete and consistent whole. (von Humboldt 1792)

Habermas's theory, as in much Western writing on ethics and justice, promotes the principle of autonomy. However, this term, with its stress on individual rights, has several meanings in the West.

For Aristotle, given that civic action was already the province of the elite, it could be argued that autonomy for his citizens of the *polis* (e.g. as opposed to slaves) was an attribute that did not require any attention.

Feinberg (1989) has claimed that there are at least four different meanings of 'autonomy' in moral and political philosophy: the capacity to govern oneself, the actual condition of self-government, a personal ideal, and a set of rights expressive of one's sovereignty over oneself. The last one is a useful notion of autonomy that in more general terms might suggest self-rule, presumably achieved through self-awareness, the emancipatory state necessary for Habermasian flexibility in discourse, as well as through rationality.

There is also the Kantian moral value of autonomy, from which Habermas takes some of his ideas, and two points can be made here.

Firstly, the Kantian notion of autonomy is often thought to be highly individualistic in its ethics of responsible action. Neumann argues though that Kant was not such a champion of individualism as is commonly thought. Kant was as a champion of rationality in Neumann's (2000) account, valuing a universal trait in the human being, admiring 'rational selves in all their sameness, in their unvarying conformity to the universal principles of pure practical reason'. If we take Kant's law of universalisation – that we should act as though one's actions could be universal laws – and apply it to this issue, what is universalisable is derived from our respect of the person as rational and as an end in himself/herself (Gregor 1996).

Secondly, Kant offers a theory of 'attraction and repulsion' as two forces necessary for the human, moral world. Repulsion, being 'the body filling its own space', implies individual freedom, but is counterbalanced by a strong relationship to the other – attraction is that which 'binds into unity' (Schneewind 1998). Kant implies a dialogical autonomy where care for self is complemented by care for others.

In considering these two ideas, the statement that Habermas derives some of his views on autonomy from Kant becomes more complicated than might initially be thought. As Korsgaard (1996) states, autonomy concerns the independence and authenticity of the desires (values, emotions, etc.) that move one to act and also the capacity to impose upon ourselves, by virtue of our practical identities, obligations to act.

Autonomy suggests both reason (self-governance) and the actualisation, to return to the discussion of Habermas, of the species ethic.

In his notion of the species ethic, Habermas (2003) emphasises autonomy – a Western rather than an Eastern concept (as well as rationality, arguably also more of a Western virtue than an Eastern one). Yet he qualifies autonomy by suggesting that although public 'morality' is tied very closely to an individual commitment to one's own values, the actualisation of morality depends upon being in a community of fellow agents, through the intersubjectivity provided by being language users: 'the logos of language embodies the power of the intersubjective, which precedes and grounds the subjectivity of speakers'. Thus, there seems to be a (positive or negative) tension between private and public, individual subjective virtues and intersubjective virtues (a tension that rather recalls that of the so-called Eastern values tension between the one and the whole).

It seems we may be in the right place with Habermas in terms of advising inter-locutors on how to be global dialogue agents and thus how to deal with the tension between the group and one's self – assuming that his 'power of the intersubjective' actually works.

Virtues mentioned by Asian leaders, including 'hard work, family, education, savings, and disciplined living', are certainly not alien in the Western tradition (Marshall 1994). The Western emphasis on individual human rights is not addressed in that statement. However, economist and political theorist Amartya Sen, in his 1997 lecture, 'Human Rights and Asian Values', questioned the notion that there is a distinct set of Asian values that is in tension with Western human rights thinking. By contrast, he argued that historically, respect for human rights is not an exclu-sively Western concept. In this vein, Wong (2004) argues that Confucianism and Western rights-centred morality can be brought closer 'through the interdependence of rights and community', through defining rights as recognised on the basis of their necessity for promoting the common good, rather than as necessary for personal autonomy. This is not wildly different to Aristotle's idea that the best person is not only one who exercises virtues that benefit himself/herself but also one who acts virtuously to benefit others.

Kupperman, writing on Confucius and Aristotle, suggests that the Confucian rejection of a Western rationalist individualism still assumes ethics is based on an individual's good character and education. However, for Confucius, decision-making will rely heavily on tradition and community, whereas for Aristotle, the agent has freedom to act independently of how others might have acted before.

The relationship between *ren* and the so-called Asian ideal of harmonious col-lectivism is explained in the Confucian *Analects*, through the idea of the virtuous or exemplary ruler. Sim (2010) traces the logic as follows:

> Confucius thinks that rule by an exemplary ruler is more effective for making one virtuous than rules, for if you govern effectively, 'what need is there for killing? The excellence of the exemplary person is the wind, while that of the petty person is the grass. As the wind blows, the grass is sure to bend.

This is an image of a quiescent populace happy to be ruled (hopefully) wisely. Elsewhere, Confucius talks of how the people will regulate themselves according to their 'shame' if they do not follow the exemplary leader's model expressed through wise leadership. Such an attitude, or 'ability to be moved by the exemplary person', Sim (2010) explains,

> …begins at home with filial piety (*xiao*) toward one's parents and love for other family members mediated by ritual propriety (*li*). *Li* dictates the proper behavior for all roles… in the larger community… *li* enables the extension of love for family members, in a graduated manner, to everyone else in the community. When accomplished, one will have the highest Confucian virtue of humaneness (*ren*) which includes all particular virtues like courage (*yong*), wisdom (*zhi*), appropriateness in actions (*yi*) and truthfulness (*xin*), just to name a few…filial and fraternal responsibility…is the root of humaneness (*ren*).

This statement suggests that the virtue of 'proper' behaviour is governance of the self. Yet Rosemont (2004) suggests that Confucianism can imply 'not only a sense of self-governance, but a sense of the importance of nurturing self-governance in others'. This may sound patronising, though when seen in the context of mutual agency, of mutual species endeavour, it is rather less so. Arguably, dialogue is one such situation of mutual agency.

Confucian humaneness (*ren*) leads one to benefit others; as Confucius puts it, 'Persons with humaneness (*ren*) establish others in establishing themselves and promote others in promoting themselves' (Sim 2010). There are some echoes in this notion of developing others through oneself in Chan's (1969) work on the idea of 'jen':

> Confucius has an answer to harmonise the conflict between international guidelines deriving from one culture (Western) and other cultures with strong and well established moral and legal traditions (such as in Asia). His answer is … that bioethics should and must be 'in harmony but not identical'; 'in harmony as well as diversified' … The basic values underlying international ethical guidelines, such as non-malfeasance/beneficence, respect for persons and justice are not so far different from Asian values … 'ren' encapsulates non-maleficence or 'do no harm', respect for others, and the Confucian 'Do not do to others what you do not want to be done to you'… or *jen* … translated as human heartedness, benevolence, love, compassion, kindness or humanity …. The person of *jen* is the one who in 'desiring to sustain oneself, sustains others and desiring to develop oneself develop others'.[12]

In a general sense, both Aristotle and Confucius promote the notion of community duty to others – thus Sim (2010) claims that these thinkers' ethics can 'include perfect duties to others, formulated in universal laws that are enforceable'. In addition, not only is one concerned with others but for the universe as a whole:

> …the harmony and welfare of the entire community takes precedence over the interests of individuals. Such harmony and welfare is to be achieved by the leading elite engaging in a rigorous regimen of learning and self-examination, namely, by incessant self-cultivation through sincere commitments to their respective roles in family, state, and world…Humans have to show concern and empathy not only for humans but also other beings in the universe. Thus the Confucian virtue ethic is rooted in an anthropocosmic world view, in which the individual is defined in terms of his/her roles and responsibilities within a given contextual relationship to other beings in the universe. (Song 2002)

Note the reference to education (shades of Aristotle!) and to self-reflection. Confucius's follower Mencius believed that everyone has the potential to become virtuous, but that these incipient tendencies towards virtue often fail to manifest in situations where they probably should; thus we must 'extend' the manifestations of our virtue (van Norden 2004). Community plays a primary role in self-extension, or self-development, as the person imitates his/her parents, but another force is that of self-reflection, for 'upon reflection, the self acquires an identity as well as a power for self-transformation; the self is a duality, both a reflective subject and a reflected-upon object' (Chen 2004). (This notion of a dual identity is important for global ethics, as will be discussed in Chapter 7.)

---

[12] *Jen* has been variously translated as benevolence, perfect virtue, goodness, human heartedness, love, altruism, etc. According to Chan, none of these express all the meanings of the term. It means a particular virtue, benevolence, and a general virtue, the basis of all goodness.

Song (2002) offers an interesting, if not altogether convincing, argument that the 'anthropocosmic' nature of Confucianism is essentially a version of care ethics linked to eco-ethics, yet also suggests a dialectical self:

> Whereas modern liberalism takes as its task the systemic guarantee of individual autonomy and rights and their protection from outside coercion, Confucian anthropocosmic ethics concerns itself with the mental, spiritual effort of intellectuals, which, first and foremost starts with resolving the conflict within oneself between the '*dao*-mind' (*daoxin*, the will to observe the public good) and the 'human-mind' (*renxin*, the will to pursue private interests and/or selfish desires), and which further culminates in preserving the harmony of humans with nature in the universe.

Self-governance, in terms of managing the dual self, is all very clear within the Confucian emphasis on achieving harmony. What of disharmony, however, when individuals wish to disagree? What if the rights to speech and dissent are somewhat insecure within political systems known to be dismissive of individualism? Wong (2004) appears to suggest that the channels through which individual dissent can emerge are significant, in that they allow individual dissent while maintaining community consensus. In other words, the issue may not be one of incompatible virtues or values, but of incompatible process. Thus, Habermas's view is that we can see universal validity in *procedural* terms, as based on certain rules of argument presupposed upon 'conditions of rational speech, symmetry, and reciprocal recognition' (Kelly 1990). However, one is reminded of Engelhardt's (2006) belief in a 'collapse of consensus' on bioethical issues demonstrated by passionate and persistent disagreement that cannot be resolved through reasoning. Consensus is never easy.

To return to the problem as originally stated, Eastern and Western virtues differ vastly on this issue of individualism. Sandel (1996) has stated that the overemphasis on choosing our own aims in the Western liberal tradition may result in our denial that 'we can ever be claimed by ends we have not chosen – ends given by nature or God, for example, or by our identities as members of families, peoples, cultures, or traditions' (p. 70). In other words, we may overstress individualistic virtues at the expense of our very real ties to our communities. In this vein, Widdows (2011) notes that:

> The 'western' moral agent is an autonomous, isolated, free, choosing individual and the Asian moral agent is a connected, community-defined, relational being …. This division is false, of course. Western individuals are not isolated beings making choices in a vacuum. To present human beings as making judgments outside their culture and background is to ignore the historically and socially constructed nature of human beings. The eastern picture is no better – that of an amalgamated creature, conjoined to relations and the family with no distinguishable personhood or identity. Such a person would be entirely passive and lack any sense of self, preference, decision-making and the ability to form relationships – again, not a realistic picture of a human being.

Perhaps Amartya Sen (1998) has the last word on this belief that Eastern values are utterly communitarian:

> There is much variety in Asian intellectual traditions, and many writers did emphasize the importance of freedom and tolerance, and some even saw this as the entitlement of every human being. The language of freedom is very important, for example, in Buddhism, which

originated and first flourished in South Asia and then spread to Southeast Asia and East Asia, including China, Japan, Korea, and Thailand. In this context it is important to recognize that Buddhist philosophy not only emphasized freedom as a form of life but also gave it a political content. To give just one example, the Indian emperor Ashoka in the third century BCE presented many political inscriptions in favor of tolerance and individual freedom, both as a part of state policy and in the relation of different people to each other. The domain of toleration, Ashoka argued, must include everybody without exception.

In conclusion, the Asian (Confucian) virtuous agent is similar to the Aristotelian and Habermasian one in that:

- Both saw moral virtues as derived not from a universal moral calculus but from a careful process of personal discovery.
- Both have an intersubjective view of ethics based on a dialectical relationship between self and others (Table 6.1).

Table 6.1 Aristotelian, Habermasian, and Confucian 'virtues'

	Aristotelian virtues	Habermasian virtues	Confucian virtues
What is a virtue?	A quality which enables a person to achieve, through right choice, a good and noble life of happiness and prosperity	Willingness to communicate to achieve consensus – part of the 'species ethic' and the practical skills required to realize that ethic	That which promotes social harmony
How virtue is achieved	Through habituation and moderation or temperance, good temper, patience	Through rational discourse and self-realisation, self-reflection	Through imitation and self-reflection
Chief virtue	Practical reason (*phronesis*)	Reason, which allows the validity of statements to be tested and thus agreed upon	*Ren*, or benevolence, which allows a meaningful pursuit of *li*, or propriety/order
Expressed through...	Deliberative wisdom	The ability to deal with complex information in an insightful and wise way, flexibility; self-reflective knowledge, or sincerity	Correct and caring social behaviour
Communicates	With truthfulness and impartiality	Intelligibly and sincerely	With a view to harmony
Is aimed at:	Greater good (thus, altruism) – including generosity, a high-minded sense of justice, and proper ambition	Consensus	Harmony (*tao*)
Is similar in:	terms of emphasis on the self	On the intersubjective self	On the self within the collective

The notion of setting up a discourse process that allows individual self-discovery, and an emancipatory dialogue, may seem somewhat unfeasible, yet when the Habermasian notion of intersubjectivity is explored further, it becomes less so. It is the core mechanism through which emancipatory knowledge, and thus useful discourse outcomes, might be achieved. This may go some way towards resolving the problems with which we began this chapter – namely, that of achieving compromise for the good of the whole rather than merely the self; and, conversely, of ensuring the self has enough autonomy to speak sincerely within the dialogue process.

6.6 Intersubjectivity

> The farther individuation progresses, the more the individual subject is caught up in a... network of reciprocal dependencies (Habermas 1989–1990).
>
>   The question: "Can Complex Societies Form a Rational Identity?" already indicates how I wish to use the term 'identity.' A society does not just have an identity ascribed to it in the trivial sense an object does, which can be identified by various observers as being the same 'thing,' although they may apprehend and describe it in different ways. In a certain sense a society achieves or, let me say, produces its identity; and it is by virtue of its own efforts that it does not lose it. To speak, moreover, of the 'rational' identity of society reveals that the concept has a normative content. (Habermas 1974)

As Gray and Lovat (2007) argue, implicit in Habermas's emancipatory knowing is the idea of the 'self-reflective knower' going beyond self-knowledge to take a 'stand for justice ... that spills over into practical action'. This implies a quality or virtue of transcending one's own limited viewpoint and interest, to become aware of a greater issue, or another person (Hendley 2000). It may seem idealistic; no wonder that Habermas once phrased the idea of transcending one's own viewpoint through a religious image, suggesting 'God' as the name for a 'communication structure that forces men, on pain of a loss of their humanity, to go beyond their accidental, empirical nature to encounter one another indirectly, that is, across an objective something that they themselves are not (Habermas 1992). How does one achieve this transcendence? Or perhaps one might call it a virtuous identity, in that it implies both self-realisation and an acceptance of others' beliefs in a sincere and rational moment of insight. The important thing in such a moment is that it demonstrates not only the exercise of individual reason and judgement but of *mutual* (i.e. consensual) reason and judgement.

A virtuous identity arises out of a network of relationships. This does not mean forgetting or downplaying the needs of the self. Habermas argues that one's 'centre' or self is created partly through externalisation in 'communicatively produced interpersonal relationships. The self should be strong and self-aware in a 'web of intersubjective relations of mutual recognition by which individuals survive as members of a community' (Habermas 1989–1990). Habermas argues that discourse requires strong individual contributions to discussion so that the participants become collectively convinced of the validity of a norm. Thus, the notion of intersubjectivity relies on a strong and individual self.

Habermas's (1991a, b) virtue of 'considerateness' is key, for it 'has the twofold objective of defending the integrity of the individual and of preserving the vital fabric of ties of mutual recognition through which individuals reciprocally stabilize their fragile identities'. In fact, Habermas (1993) sees the strength of the self and the strength of the community/other as indispensably symbiotic:

> The person develops an inner life and achieves a stable identity only to the extent that he also externalizes himself in communicatively generated interpersonal relations and implicates himself in an ever denser and more differentiated network of reciprocal vulnerabilities, thereby rendering himself in need of protection. From this anthropological point of view, morality can be conceived as the protective institution that compensates for a constitutional precariousness implicit in the social or cultural form of life itself. Moral institutions tell us how we should behave towards one another to counteract the extreme vulnerability of the individual through protection and considerateness. Nobody can preserve his integrity by himself alone ... Morality is aimed at the chronic susceptibility of personal integrity implicit in the structure of linguistically mediated interactions, which is more deep-seated than the tangible vulnerability of bodily integrity, though connected with it.

The above statement suggests a nexus between identity and the network of mutual (considerate) recognition. There is more to it than this; however, assuming that individuals deserve equal consideration, we still have to know what it means to regard the interests of others as being as important as our own. Hare (1963) thinks we should imagine ourselves in the place of others, with the interests that they have, before weighing these interests in relation to each other to determine which course of action would maximise their satisfaction. In a sense, we are adopting the stance of the other; as this is difficult though, we might consider the adoption of a 'public' role, which means considering the good of the whole as opposed to merely the good of the self. Thus, Habermas's (1989–1990) view of solidarity is 'rooted in the realization that each person must take responsibility for the other, because as coassociates all must have an interest in the integrity of their shared life context in the same way'.

Solidarity to Habermas means the coordination and pursuit of individual (or joint) goals on the basis of a shared understanding that the goals are inherently reasonable or merit worthy. This gets around the so-called flaw in the Rawlsian suggestion that parties might act as if 'rational and mutually disinterested' when settling an issue – it is by no means guaranteed that those debating an issue will argue towards a position of mutual benefit, or are motivated only by a concern for the better argument.[13] Rawls himself, noting the potential gap between acting justly and the (Kantian) desire to express the self as a free moral person, implies that the desire for reciprocity may be the necessary missing ingredient (Baldwin 2008; Rawls 1971). Rawls arguably reworks Smith's theory of sympathy, according to which, 'howsoever selfish man may be supposed, there are evidently some principles in his nature, which interest him in the fortune of others, and render their happiness necessary to him, though he derives nothing from it except the pleasure of

---

[13] Habermas's discourse ethics has been criticised for being utopian and idealistic by Foucault (1988) and Flyvberg (1998).

Table 6.2 The skill set of the (global) virtuous discourse agent

The speaker must be:	How this works	Is this universal?
Motivated by the species ethic, i.e. have a desire to communicate for the greater good of consensus	Emotion and the species ethic provide universal impulses towards communication	Yes
Must be able to compromise, be flexible (be swayed by reason)	Reason is part of the process; a sincere considerateness is brought to that process of reasoning	*Ren* ethics and care ethics have much in common; however, reason is seen as less important in the East
Must possess a practical desire and ability to act	A function of practical discourse ethics, particularly in the form of pTA	A function of practical discourse ethics, particularly in the form of pTA
Must act wisely, i.e. rationally (neutrally, temperately, consistently) in terms of understanding one's own motivations and arguments, and those of others – in short, must be self-aware	Self-awareness is developed through dialogue and dialectic – just as awareness of the other is developed. Consensus is the desired result	Awareness of the other is developed through dialogue – harmony is seen as more important in the East
Must be able to act autonomously and justly	Intersubjectivity implies that the autonomous self requires relationality	The community-minded person seeks harmony of self and whole

seeing it' (Smith 1759).[14] Communicative reason is the practice of solidarity (Pensky 2008).

It seems that of all the virtues, the only ones that particularly matter after all are the ones of solidarity and self-knowledge, existing within a dual structure. In other words, we have a global value (solidarity) and a private/personal one (self-knowledge) (Table 6.2).

6.7 Global or Universal Virtues

> The cultural diversities of the world are there to stay but human beings, regardless of their cultural traditions, do share some universal values, such as love and compassion, doing no harm… These values are the foundation of a global bioethics which should be independent of the norms of any particular culture because foundational moral values transcend particular cultural values. (De Bary 1996)[15]

---

[14] See Library of Economics and Liberty (2000) The Theory of Moral Sentiments. http://www. econlib.org/library/Smith/smMS.html

[15] See also Tai (2007a, b).

Before looking at identity formation as a way towards global agency in ethics, it is worth asking whether there is any particular agreement as yet on global values and whether East and West have any similarities in their general concept of identity.

As Widdows and Sandel suggest, the individual and the collective are not quite as distanced from each other as we might initially expect. Kupperman (2004) leaves the door open for a compromise position when he looks at the issue of the unity of the self and argues that Confucius may have believed in a 'self-as-collage' approach. What does this mean? De Castro (1999) refers to collage in a more useful fashion when discussing universalisation that 'can be understood as a process of putting together a collage of varying perspectives without having to assert a neutral standard of measure'. 'Intersections of experience', as he terms them, may lead to that universalising moment. Arguing that the main issue is 'one of identity', he suggests that authenticity (being true to one's self and to the values that constitute our unique cultural perspective) is the key to ethics (De Castro 1999).

There is a simpler way of viewing this balance between individual and collective. Empathy – or to use Habermas's term, considerateness – may indeed be a universal virtue; Western care ethics is close to Confucian *ren* (or *jen*) ethics, in which the central principle, *ren*, means 'love and care for others'. The focus is put on *inter*dependent human relationships; relational ties suggest the basis for interdependence, whether within or across communities (Hui 2000). Kishore (2003) suggests virtues such as 'love, trust, righteousness, compassion, tolerance, fairness, forgiveness, beneficence, sacrifice, and concern for the weak', while Tai (2007a, b) presents virtues and principles based on Chinese traditions of 'compassion, respect, righteousness, responsibility, and *ahimsa*' (the latter meaning the sacredness of all living things).

Sass (2009) argues that we need a radical paradigm shift towards a new and healthier way of caring. Referring to health-care ethics, he argues that compassion and communication are vital; 'also the competence and cooperation of stakeholders involved will be essential for a new culture in the cultivation of human-human interactions'. The five golden C-principles are, therefore, 'competence, compassion, communication, cooperation, and culture'.[16]

Gould (1988) asserts the relationality and connectedness of human beings, which militates against a conception of human beings as isolated, separate individuals, seeing us as 'individuals in relations', i.e. individualism being tempered as all activity includes the recognition of others, their needs, and the individual's relationship with them.[17]

The individualistic concept of autonomy might be replaced by the notion of 'holistic harmony'. The word 'holistic', though somewhat overused, is a useful reminder, in that it implies balancing various needs within the community. Thus, Hongladarom, looking at how technology might be adapted to Thai Buddhist concepts of emptiness and compassion, suggests that nanotechnology would need to be

---

[16] See also Sass (2011).

[17] See also Gould (2004).

geared more towards public good than personal gain. It should be accompanied by
a concern for its impact on others:

> ...the Buddhist attitude would be to say that the primary motivation behind such introduc-
> tion of the products should not be exclusively for the personal gain of the owners or the
> shareholders of the firms themselves, but primarily for the benefits of all who will find a use
> for the products. (Hongladarom 2009)

In 2005, Knappers and Chadwick noted a new trend in ethics. They found that
there has been a movement away from prioritising the dominant Western set of
biomedical principles of autonomy, non-maleficence, beneficence, and justice,
towards those of 'reciprocity, mutuality, solidarity, citizenry, and universality'
(Knappers and Chadwick 1994).[18] These are not only less teleological and more
process oriented (Salter and Salter 2007) but also suggest an Aristotelian balance
between the common good and autonomy, a common good being 'proper to and
attainable only by the community, yet individually shared by its members' (Aristotle
1998). Such a balance would seem, in terms of the differing global bioethics values
that have been identified, to suggest a meeting between Western individualism and
Eastern or African communalism. If the term 'good' seems perhaps too culturally
loaded, then one might use Sen's term 'flourishing' – a nice, practical term and one
more easily quantifiable. As Radha Krisha (2010) argues in the context of Malaysian
bioethics:

> The ethical landscape of modern multicultural, multiracial, multi-faith Malaysia ... may
> indeed have been what Ladikas [and Schroeder] envisaged when describing global ethics as
> 'the attempt to agree on fundamental conditions for human flourishing and to secure them
> for all.

In conclusion, if we return to Habermas, it is less important that the participant
in pTA or discourse possesses particular virtues, compared with virtuous skills that
enable dialogue. The most crucial aspect of dialogue may be that it is in fact dialogue; the perception that cultural, national, and racial identity is being subverted by
a colonising Western form of ethics may be the greatest bar to such a debate.

Thus, Sen's (1996) concern that 'the lives of Asians – their beliefs and traditions,
their rules and regulations, their achievements and failures, and ultimately their
lives and freedoms' can appear to be already decided.

Western dialogue may be broadly defined as teleological; Asian dialogue appears,
again speaking broadly, to be relational (Habermas 2004).

What we still need to understand is the procedural principle of universalisation
that Habermas brings to the discourse model and that allows for the intersubjective
recognition of common goals. This will be discussed in the next chapter, together
with the issue of identity formation.

Thus, when asking the question, are virtuous discourse agents the same in the
East and the West, the argument is that Habermasian thought (particularly as developed by people like Benhabib) incorporates an element of care ethics. Thus, there is

---

[18] A European group of bioethicists, financed by the EU, has suggested 'autonomy, dignity, integrity, and vulnerability' as a specific European list.

a point of crossover with Confucian *ren* (benevolence or kindness) ethics. Habermas's view of emotion relates to procedural issues such as coming to dialogue with a sincere and committed mindset, one that allows for empathy or 'considerateness'. *Ren* ethics seems fairly similar to the Habermasian species ethic on the following points:

1. It reveals itself in a relational form.
2. It requires sincerity or 'inner form'.
3. It is linked to ethical maturation – a concept Habermas has discussed in his view of human development and how the human being progresses through various stages of knowledge.

However, even if there is a similarity in the agent's approach to discourse, this is not to say that his or her 'knowledge form', i.e. mode of argumentation, is alike. The main issue appears to be that despite having comparable virtues in the East and the West, there is still a problem in determining how global virtuous agents might approach dialogue. This requires us to look further at the idea of agent identity formation.

References

Ames RT, Rosemont H Jr (1999) The analects of Confucius: a philosophical translation. Ballantine Books, New York

Annas J (1993) The morality of happiness. Oxford University Press, New York

Apel KO (2008) Globalisation and the need for universal ethics. In: Cortina A, Garcia-Marquez D, Connill J (eds) Public reason and applied ethics. The ways of practical reason in a pluralist society. Ashgate, Aldershot, pp 135–154

Aristotle (1998) The Nichomachean ethics. Oxford University Press, Oxford

Attfield R (1995) Value, obligation and meta-ethics. Rodopi, Amsterdam

Bakhtin M (1981) The dialogic imagination. Four essays. University of Texas Press, Austin

Bakhtin M (1984) Problems of Dostoevsky's poetics. University of Minnesota Press, Minneapolis

Bakhtin M (1986) Speech genres and other late essays. University of Texas Press, Austin

Bakhtin M (1999) Problems of Dostoevsky's poetics. University of Minnesota Press, Minneapolis

Baldwin T (2008) Rawls and moral psychology. In: Oxford studies in metaethics, vol 3. Oxford University Press, Oxford, pp 247–270

Baxter LA (1988) A dialectical perspective of communication strategies in relationship development. In: Duck S (ed) Handbook of personal relationships. Wiley, New York, pp 257–273

Blaug R (2000) Citizenship and political judgement: between discourse ethics and phronesis. Res Publica 6:179–198

Bohman J, Rehg W (2014) Stanford encyclopedia of philosophy. Jürgen Habermas. http://plato.stanford.edu/entries/habermas/#HabDisThe. Accessed 29 Dec 2014

Borradori G (2003) Philosophy in a time of terror, dialogues with Jürgen Habermas and Jacques Derrida. http://www.press.uchicago.edu/Misc/Chicago/066649.html. Accessed 29 Dec 2014

Braaten J (1991) Habermas's critical theory of society. SUNY Press, New York

Brandt R (1995) The definition of an "Ideal observer" theory in ethics. Philos Phenomenol Res 15:407–413

Brentano F (1973) The foundation and construction of ethics. Humanities Press, New York

Broadie S (1991) Ethics with Aristotle. Oxford University Press, New York

Broodryk J (2007) Understanding South Africa: the uBuntu way of living. UBuntu School of Philosophy, Waterkloof

Burnor R, Raley Y (2011) Ethical choices. An introduction to moral philosophy with cases. Oxford University Press, Oxford

Chan WT (1969) A source book in Chinese philosophy. Princeton University Press, Princeton

Chen C (2004) Theory of Confucian selfhood. In: Shun K, Wong DB (eds) Confucian ethics. A comparative study of self, autonomy and community. Cambridge University Press, Cambridge, pp 124–148

Ch'ung W, Zhen F (2003) The budding of the structural perspective in China (i.e. in the Eastern Ecumene). http://www.oocities.org/theophoretos/fanzhen.html. Accessed 29 Dec 2014

Conley S (1988) Flourishing and the failure of the ethics of virtue. Midwest Stud Philos 13(1):83–96

Cortina A (2008) The public task of applied ethics: transnational civic ethics. In: Cortina A, Garcia-Marquez D, Connill J (eds) Public reason and applied ethics. The ways of practical reason in a pluralist society. Ashgate, Aldershot, pp 9–32

Dare T (2002) Applied ethics, challenges to. In: Chadwick R, Schroeder D (eds) Applied ethics. Critical concepts in philosophy. Routledge, London, pp 23–35

De Bary TW (1996) The Buddhist tradition in India, China and Japan. The Modern Library, New York

De Castro LD (1999) Is there an Asian bioethics? Bioethics 13:227–235

Donohoe J (2004) Husserl on ethics and intersubjectivity. From static to genetic phenomenology. Humanity Books, Amherst

Dostovesky F (1862) Notes from underground. https://ebooks.adelaide.edu.au/d/dostoyevsky/d72n/chapter9.html. Accessed 29 Dec 2014

Engelhardt HT Jr (ed) (2006) Global bioethics: the collapse of consensus. M&M Scrivener Press, Salem

Enslin P, Horstheke K (2004) Can Ubuntu provide a model for citizenship education in African democracies? Comp Educ 40(4):545–558

Eze MO (2010) Intellectual history in contemporary South Africa. Palgrave Macmillan, New York

Feinberg J (1989) Autonomy in the inner citadel: essays on individual autonomy. Oxford University Press, New York

Finlayson JG (2005) Habermas. A very short introduction. Oxford University Press, Oxford

Flyvbjerg B (1998) Habermas and Foucault: thinkers for civil society? Br J Sociol 49(2):208–233

Foucault M (1988) The ethic of care for the self as a practice of freedom. In: Bernauer J, Rasmussen D (eds) The final Foucault. MIT Press, Cambridge, pp 1–20

Gichure CW (2006) Teaching business ethics in Africa: what ethical orientation? The case of East and Central Africa. J Bus Ethics 63:39–52

Glenna LL (2008) Redeeming labor: making explicit the virtue theory in Habermas's discourse ethics. Crit Sociol 34(6):767–786

Gould C (1988) Rethinking democracy: freedom and social cooperation in politics, economy, and state. Cambridge University Press, Cambridge

Gould C (2004) Globalising democracy and human rights. Cambridge University Press, Cambridge

Gray M, Lovat T (2007) Horse and carriage: why Habermas's discourse ethics gives virtue a praxis in social work. Ethics Soc Welf 1(3):310–328

Gregor M (ed) (1996) The metaphysics of morals. Cambridge University Press, Cambridge

Habermas J (1974) On social identity. Telos 19:91–103

Habermas J (1979) What is universal pragmatics? In: Habermas J (ed) Communication and the evolution of society. Beacon, Boston, pp 1–69

Habermas J (1989–1980) Justice and solidarity: on the discussion concerning "Stage 6". Philos Forum 31(Fall–Winter):32–52

Habermas J (1991a) Moral consciousness and communicative action. MIT Press, Cambridge

Habermas J (1991b) A reply. In: Honneth A, Joas H (eds) Communicative action. Essays on Jürgen Habermas' the theory of communicative action. MIT Press, Cambridge, pp 220–246

Habermas J (1992) Postmetaphysical thinking: philosophical essays. MIT Press, Cambridge

Habermas J (1993) Justification and application: remarks on discourse ethics. MIT Press, Cambridge, MA

Habermas J (1998) Between facts and norms. MIT Press, Cambridge, MA

Habermas J (2001) On the pragmatics of social interaction. MIT Press, Cambridge, MA

Habermas J (2003) The future of human nature. Polity, Cambridge

Habermas J (2004) The theory of communicative action. MIT Press, Cambridge

Hare R (1963) Freedom and reason. Oxford University Press, Oxford

Hendley S (2000) From communicative action to the face of the other. Levinas and Habermas on language, obligation, and community. Lexington Books, Lanham

Hongladarom S (2009) Nanotechnology, development and Buddhist values. NanoEthics 3:97–107

Hui E (2000) Jen and perichoresis: the Confucian and Christian bases of the relational person. In: Becker GK (ed) The moral status of persons: perspectives on bioethics. Rodopi, Atlanta, pp 95–118

Hui E (2004) Personhood and bioethics: Chinese perspective. In: Qiu R (ed) Bioethics – an Asian perspective: a quest for moral diversity. Springer, Dordrecht

Hursthouse R (1999) On virtue ethics. Oxford University Press, Oxford

Hursthouse R (2011) What does the Aristotelian phronimos know? In: Jost L, Wuerth J (eds) Perfecting virtue. New essays on Kantian ethics and virtue ethics. Cambridge University Press, Cambridge, pp 38–57

Hutchings K (2010) Global ethics. An introduction. Polity, Cambridge

Iser M (2003) Habermas on virtue. https://www.bu.edu/wcp/Papers/Cont/ContIser.htm. Accessed 29 Dec 2014

Kawall J (2006) On the moral epistemology of ideal observer theories. Ethic Theory Moral Pract 9(3):359–374

Keller M, Edelstein W, Krettenauer T, Faug F, Ge F (2005) Reasoning about moral obligations and interpersonal responsibilities in different cultural contexts. In: Edelstein W, Nunner-Winkler G (eds) Morality in context. Elsevier, Amsterdam, pp 317–337

Kelly M (1990) The Gadamer-Habermas debate revisited: the question of ethics. In: Rasmussen D (ed) Universalism vs. communitarianism. Contemporary debates in ethics. MIT Press, Cambridge, pp 139–162

Kirchengast U (2011) Solomon on the role of virtue ethics in business. In: Dierksmeier C, Amann W, von Kimakowitz E, Spitzek H, Pirson M (eds) Humanistic ethics in the age of globality. Palgrave Macmillan, Basingstoke, pp 187–209

Kishore RR (2003) End of life issues and moral certainty. A discovery through Hinduism. Eubios J Asian Int Bioeth 13:210–213

Knappers ABM, Chadwick R (1994) The human genome project: under an international ethical microscope. Science 264:2035–2036

Korsgaard C (1986) Skepticism about practical reason. J Philos 83(1):5–25

Korsgaard C (1996) The sources of normativity. Cambridge University Press, Cambridge

Korsgaard C (2008) The constitution of agency. Essays on practical reason and moral psychology. Oxford University Press, Oxford

Kresse K (2011) African humanism and case study from Swahili coast. In: Dierksmeier C, Amann W, von Kimakowitz E, Spitzek H, Pirson M (eds) Humanistic ethics in the age of globality. Palgrave Macmillan, Basingstoke, pp 246–266

Kupperman JJ (2004) Tradition and community in the formation of character and self. In: Shun K, Wong DB (eds) Confucian ethics. A comparative study of self, autonomy and community. Cambridge University Press, Cambridge, pp 103–123

Li P (1999) Confucian values and personal health. In: Fan R (ed) Confucian bioethics. Kluwer Academic Publishers, Hingham, pp 24–48

Lovat T (2003) The contribution of proportionism to bioethical deliberation in a moderately post-scientific age. http://www.mcauley.acu.edu.au/theology/Issue3/index.html. Accessed 29 Dec 2014

Lovat T, Gray M (2008) Towards a proportionist social work ethics: a Habermasian perspective. Br J Soc Work 38(6):1100–1111

Luhabe W (2002) Defining moments: experiences of black executives in South Africa's work-
    place. University of Natal Press, Pietermaritzburg
Lutz DW (2009) African Ubuntu philosophy and global management. J Bus Ethics 84:313–328
Lyman D (1856) The moral sayings of Publius Syrus, a Roman slave. LE Barnard & Company,
    Boston
MacIntyre A (1981) After virtue. A study in moral theory. Duckworth, London
Mackie JL (1976) Problems from Locke. Clarendon, Oxford
Marshall P (1994) Bad company: Western values criticized in Asia. Areopagus 7(4):11
Meilaender GC (1984) The theory and practice of virtue. University of Notre Dame Press, Notre
    Dame
Mendieta E (2002) The adventures of transcendental philosophy. In: Apel KO (ed) Semiotics and
    discourse ethics. Rowman & Littlefield, Lanham
Neumann M (2000) Did Kant respect persons? Res Publica 6:285–299
Nussbaum M (1992) Human functioning and social justice: in defense of Aristotelian essentialism.
    Polit Theory 20(2). http://ptx.sagepub.com/content/20/2/202.short. Accessed 29 Dec 2014
Nuyen AT (2006) The self and its virtues: is there a Chinese–Western contrast? In: Chon K, Liu Y
    (eds) Conceptions of virtue: East and West. Marshall Cavendish, Singapore, pp 237–254
Orwell G, Gessen K (2008) All art is propaganda. Critical essays. Houghton Mifflin Harcourt,
    London
Pensky M (2008) The ends of solidarity. Discourse theory in ethics and politics. State University
    of New York Press, New York
Plato (2008) The Republic. http://www.gutenberg.org/files/1497/1497-h/1497-h.htm. Accessed 29
    Dec 2014
Radha Krishna LK (2010) Global ethics – a Malaysia – Singaporean perspective. Eubios J Asian
    Int Bioeth 20(September):140–146
Ramose MB (1999) African philosophy through Ubuntu. Mond Books, Harare
Rawlins WK (1988) A dialectical analysis of the tensions, functions and strategic challenges of
    communication in young adult friendships. In: Anderson JA (ed) Communication yearbook 12.
    Sage, Newbury, pp 157–189
Rawls J (1971) A theory of justice. Belknap Press of Harvard University Press, Cambridge
Rehg W (1994) Insight and solidarity. A study in the discourse ethics of Jürgen Habermas.
    University of California Press, Berkeley
Rosemont H Jr (2004) Whose democracy? Which rights? A Confucian critique of modern Western
    liberalism. In: Shun K, Wong DB (eds) Confucian ethics. A comparative study of self, auton-
    omy and community. Cambridge University Press, Cambridge, pp 49–71
Salter B, Salter C (2007) Bioethics and the global moral economy: the cultural politics of human
    embryonic stem cell science. Sci Technol Hum Value 32:554–581
Sandel M (1996) Democracy's discontent. Harvard University Press, Cambridge
Sandler RL (2007) Character and environment. Columbia University Press, New York
Sass HM (2009) Asian and European roots of bioethics: Fritz Jahr's 1927 definition and vision of
    bioethics. Asian Rev Bioeth 1(3):185–197
Sass HM (2011) Cultivating and harmonising virtues and principles. Asian Rev Bioeth 3(1):36–47
Schneewind JB (1998) The invention of autonomy: a history of modern moral philosophy.
    Cambridge University Press, Cambridge
Sen A (1996) Thinking about human rights and Asian values. http://www.carnegiecouncil.org/
    publications/archive/dialogue/1_04/articles/519.html. Accessed 29 Dec 2014
Sen A (1997) Human rights and Asian values. Sixteenth Morgenthau memorial lecture on ethics &
    foreign policy. https://www.carnegiecouncil.org/publications/archive/morgenthau/254.html/_
    res/id=sa_File1/254_sen.pdf. Accessed 29 Dec 2014
Sen A (1998) Universal truths: human rights and the westernizing illusion. Harv Int Rev 20(3
    Summer):40–43. https://www.mtholyoke.edu/acad/intrel/asian%20values/sen.htm. Accessed
    29 Dec 2014
Sherman N (1999) Aristotle's ethics. Rowman & Littlefield, London

Shutte A (1993) Ubuntu: an ethic for the new south Africa. Cluster Publications, Cape Town

Siep L (2005) Virtues, values, and moral objectivity. In: Gill C (ed) Issues in ancient and modern ethics. Clarendon, Oxford, pp 83–98

Sim M (2010) Rethinking virtue ethics and social justice with Aristotle and Confucius. Asian Philos 20(2):195–213

Sithole J (2001) Africa can only use own culture to influence globalization. http://www.globalpolicy.org/globaliz/cultural/2001/0515afr.htm. Accessed 29 Dec 2014

Slote M (1983) Goods and virtues. Oxford University Press, Oxford

Smith A (1759) The theory of moral sentiments. http://www.econlib.org/library/Smith/smMS.html. Accessed 28 Dec 2014

Solomon RC (1993) The passions: emotions and the meaning of life. Hackett, Indianapolis

Solomon RC (1995) A passion for justice. Emotions and the origins of the social contract. Rowman & Littlefield, London

Solomon RC (1999) The joy of philosophy. Thinking thin versus the passionate life. Oxford University Press, New York

Solomon RC (2003) Not passion's slave. Emotions and choice. Oxford University Press, Oxford

Song Y (2002) Crisis of cultural identity in East Asia: on the meaning of Confucian ethics in the age of globalisation. Asian Philos 12(2):109–125

Stansbury J (2009) Reasoned moral agreement: applying discourse ethics within organizations. Bus Ethics Q 19(1):33–56

Swanton C (2003) Virtue ethics: a pluralistic view. Oxford University Press, Oxford

Swanton C (2007) Role ethics and business ethics. In: Walker RL, Ivanhoe PJ (eds) Working virtue: virtue ethics and contemporary moral problems. Oxford University Press, Oxford

Swift A (2006) Political philosophy, Part 1: Social justice. Polity, Cambridge, pp 21–49

Tai MC (2007a) The way of Asian bioethics. Princeton International Publications, Taipei

Tai MC (2007b) A Confucian perspective on bioethical principles in ethics consultation. Clin Ethics 2(4):201–207

Tambulasi R, Kayuni H (2005) Can African feet divorce western shoes? The case of 'Ubuntu' and democratic good governance in Malawi. Nord J Afr Stud 14(2):147–161

Taylor C (1989) Sources of the self. Harvard University Press, Cambridge

Thomson J (1953) The ethics of Aristotle. George Allen, London

Tucker J (2013) Stanford encyclopedia of philosophy. Jpn Confucian Philos. http://plato.stanford.edu/entries/japanese-confucian/#ConModJap. Accessed 29 Dec 2014

Tutu D (1999) No future without forgiveness. Rider, London

Van Norden BW (2004) The virtue of righteousness in Mencius. In: Shun K, Wong DB (eds) Confucian ethics. A comparative study of self, autonomy and community. Cambridge University Press, Cambridge, pp 149–182

Van Norden BW (2007) Virtue ethics and consequentialism in early Chinese philosophy. Cambridge University Press, Cambridge

Veatch RM (2004) Common morality and human finitude: a foundation for bioethics. In: Baumann E, Brink A, May A (eds) Weltanschauliche Offenheit in der Bioethik. Duncker & Humboldt, Berlin, pp 37–50

von Humboldt W (1792) The limits of state action. http://classicliberal.tripod.com/humboldt/. Accessed 29 Dec 2014

Walker MU (2003) Moral contexts. Rowman & Littlefield, London

Walker RL, Ivanhoe PJ (eds) (2007) Working virtue: virtue ethics and contemporary moral problems. Oxford University Press, Oxford

Wenstop F (2005) Mindsets, rationality and motion in multi-criteria decision analysis. J Multi-Criteria Decis Anal 13:161–172

Whetstone JT (2001) How virtue fits within business ethics. J Bus Ethics 33(2):101–114

Widdows H (2011) Western and Eastern principles and globalised bioethics. Asian Bioeth Rev 3(1):14–22

Williams B (1985) Ethics and the limits of philosophy. Harvard University Press, Cambridge

Wiredu K (1996) Cultural universals and particulars: an African perspective. Indiana University Press, Bloomington

Wong DB (2004) Rights and community in Confucianism. In: Shun K, Wong DB (eds) Confucian ethics. A comparative study of self, autonomy and community. Cambridge University Press, Cambridge, pp 31–48

Yee CS (1999) Disputes on the one thread of Chung-Shu. J Chin Philos 26(2):165–186

Chapter 7
Universalism Versus Relativism

> One who knows himself and others will find out here that East
> and West are no longer separable ('Wer sich selbst und andere
> kennt wird auch hier erkennen; Orient und Okzident sind nich
> mehr zu trennen, Sinnig zwischen beiden Welten sich zu wiegen
> lass ich gelten: Also zwischen Ost-und Westen sich bewegen sei
> zum Besten') (Goethe 1972).

> ... universalisation ... can be understood as a process of
> putting together a collage of varying perspectives ...
> 'Intersections of experience'. . . may lead to that universalising
> moment (De Castro 1999).

Abstract This chapter examines issues arising from the universalist and relativist approaches. This battle between the universal and the local still presents a challenge for policymakers developing an inclusive global and ethical approach to technology. Can participants in a Habermasian-inspired discourse/pTA achieve a 'dialogic form' or 'dialogic self', a concept informed to a degree by Korsgaard in her work on agency and self-constitution, Appiah on the dialogically constituted self, Gutmann and Thompson on first- and second-order values, and Charles Taylor on recognition and difference? As Taylor suggests, identity is created dialogically through our interactions with others. Through such intersubjectivity, we develop new vocabularies of comparison, discover shared aspects of identity, and achieve, hopefully, an 'intercultural personhood' that may lead to a 'fusion of horizons' and, so, perhaps, to a global ethics.

Keywords Universalism • Relativism • Intersubjectivity • Intercultural personhood • Agency • Self-constitution

With the formation of the IMF and the World Bank in 1944, the UN in 1945 and its subsidiary the World Health Organization in 1948, along with the UN's adoption of the Universal Declaration of Human Rights in the same year, one could begin to argue for an emerging sense of transnational moral responsibility. The ideals promoted by the new global organisations – global democracy, global economic growth and stability, global peace, and global health – suggested shared or global concerns.

© Springer International Publishing Switzerland 2015   175
S. Dalton-Brown, *Nanotechnology and Ethical Governance in the European
Union and China*, DOI 10.1007/978-3-319-18233-9_7

However, the word *global* does not always evoke positive associations of international moral responsibility. To non-Western countries, it can mean a form of cultural imperialism, while the globalised world of international trade links and global businesses derived from the reduction of barriers to the flow of goods, services, people, and capital is leading to a global 'McWorld' culture of increased financial volatility (Stiglitz 2003; Barber 1996; Chomsky 2002). Naomi Klein (2000) has argued that instead of a global village of international harmony, we find one where 'the economic divide is widening and cultural choice narrowing'. In a homogenised global world of international corporations, globalisation might be no more than consumerism without borders. So what about a global society, 'a single human community' (Kessler 2000), social globalisation leading to the creation through social media of global (virtual) villages (Tehranian 1999; Gurtov 1994), and a new, paradoxically homogenising yet multicultural global lifestyle?[1]

What unifying values might a global society espouse? What would global ethics, international ethics, or world ethics look like?[2] We assume in what follows that values are the topic for discussion, rather than political structures (Caney 2005). I will not talk about 'a set of ranked or weighted moral standards binding on rational agents' (Philips 1994), as this leads to slippery issues such as who might create universal ethical standards, and questions about who they will be imposed on. Here Kymlicka's (2007) point – that by global ethics we mean not only standards, but a second layer of 'values' – is useful:

> (global ethics is)…a two-level phenomenon. At one level, we have a self-standing international discourse, such as human rights, that seeks to define a minimum set of standards agreeable to all. At the second level, we have a multiplicity of different ethical traditions, each of which has its own account of what more, or what else, is needed above and beyond human rights.

This chapter focuses on the problem of how to combine or to achieve consensus, given this multiplicity of different ethical traditions. It will therefore start with universalism or the attempt to assert universal values.

7.1 Problems with Universalism and Relativism

Common accusations levelled against universalism are that it is paternalistic and colonial. The cross-cultural application of a 'transnational set' of values (Mann et al. 1999) is seen as a form of neo-colonialism (Widdows 2007) and, more generally, as undermining cultural pluralism and human dignity (Andorno 2009). However, the concept of 'universalism' might also be seen in a Kantian context, as about maxims of common understanding approached in a less dictatorial manner.

---

[1] Yet, as Appiah (2006) suggests, globalisation can also threaten homogeneity.

[2] Dower (2007) prefers the term 'world ethics'. His reasoning is given in *World Ethics: The New Agenda*.

Philpott (2007) notes Pope John Paul II and President Khatemi of Iran's suggested 'dialogue between civilisations', which might enunciate values and rights 'in ways that are more consonant with the claims of multiple traditions', as a better way forward than Western-centric universalism. We might then agree with Sakamoto (1999), who argues that global ethics must be based on 'the traditional ethos of each region'.

A modern ethics must acknowledge postmodern 'globality', which 'entails a turn away both from provincial gaze and from the exotic gaze of the colonizer (and the colonized),' and works according to 'networks of global inter-communication' (Therborn 2000). It is true, as Ferraro states, that 'the new universalism will have to come to terms with one important aspect of the transformation undergone by philosophy during the last hundred years, namely, the discovery of the contextuality of knowledge and normativity' (Ferrara 1990). In other words, universalism has been in retreat before relativism, or contextuality, for some decades.

However, relativism is not entirely wonderful either. Like universalism – which can be broadly seen as an agreement on certain ethical principles, but which is often seemingly interpreted as a form of Western cultural fascism – relativism can be broadly defined as sensitivity to cultural variation, as well as a pragmatic understanding that such variation needs to be taken into account when working towards common agreement.

Taken to its furthest extreme, relativism implies Bauman's (2000) 'liquid modernity', a fluid situation in which, lacking solid frames of reference, we must create our own structures, establishing individual meaning under conditions of uncertainty, operating without semantic agreement as we 'run against the boundaries of language' (Rorty 1980), and constructing in fact our own 'languages' or worldviews, which are not always subject to external validation (Wittgenstein 1965). Thus, there can be no agreement on ethics, given that there can be no causal explanations (Nagel 1979) and no disagreements about facts, but rather disagreements about 'differences in values about which nothing can be said' (Hursthouse 1999). There are varying types of relativism (e.g. which can be discussed in the context of subjectivity or metaethics) (Brandt 2001), but most describe a 'relativist' as someone who can always find a cultural reason that requires an accommodation or exception from a general moral or behavioural rule. At its most extreme, relativism predicates a world of utter subjectivity or contextuality, leading to the conclusion that the relativist can also be accused of undermining the human dignity of others, just as the universalist does (Cook 1999). After all, one can use it to justify doing what one wishes to do anyway, with little care for others' values. As Mackie notes:

> Disagreement about moral codes seems to reflect people's adherence to and participation in different ways of life. The causal connection seems to be mainly that way round: it is that people approve of monogamy because they participate in a monogamous way of life rather than that they participate in a monogamous way of life because they approve of monogamy' (Mackie 1990).

Given that both universalism and relativism are problematic, what is a global ethicist, or policymaker, to do? One might first take comfort in the fact that the two theories are not as opposed as they might initially seem, which perhaps hints at some form of possible synthesis. In support of this claim, we might propose two notions. The first comes from Stephan Lukes, who has argued for a point of inter-section between relativism and universalism. To Lukes (2008), acknowledging the facts of moral diversity and value pluralism does 'not entail abstention from judging others (and their judging us)'. Thus, Lukes looks at the 'two questionable assump-tions' underlying contemporary multicultural thinking: first, that the world divides neatly into distinct cultures (Must cultural identities be seen as 'distinct and incom-patible'?) and second that everyone needs just one such culture to live a meaningful life. He argues that to see culture as an integrated concept rather than as a cluster (possibly even a loose one) of practices, or a set of processes, is a detrimental myth for our times and adds that:

> The loss of a widely shared worldview with secure metaphysical and religious foundations (was it really so consensual and secure?) does not render us unable to make universally applicable judgments (Lukes 2008).

In other words, one does not need an agreed universal set of values or even laws in order to decide whether something is wrong. Such a decision can be based on an individual worldview, but is also, in the Kantian sense, seen as something that should have a more general application than 'I think this is right'. Lukes' suggestion is appealing, as it implies some form of intersection between relative judgments ('I judge') and universal judgments ('we should all judge this way').

On the other side of the argument, one might also argue that universalism is never fully universal, but is rather the imposition of an idea held by the majority. Scott (1999) comments on this universalisation of a limited ethos:

> The frequent and recent criticisms of universalisation are right in their claims that this movement itself...incites something like tribal wars, now conducted as struggles for cul-tural domination and refusal of amalgamation (i.e. of transformation of identity)... The question arises from a limited ethos combined with its universalization, which transgresses its own limits, and its claim to authority.

Scott suggests that (a) a universal view is not feasible but rather we can only achieve a limited one, and (b) that ignoring such limitation while asserting the uni-versal application of a limited viewpoint is actually highly destructive. Universalisation claims are therefore quite hypocritical or reflect self-deception caused by the mistaken belief that one's views are sufficiently good to be univer-sally applied. Nussbaum has a pointed example of this kind of limited universalism, discussing attitudes to women who wear the *burqa* in a Western country. Arguments for banning the *burqa*, (effectively arguments against religious/cultural toleration), according to Nussbaum (2010), revolve around the notion that it affects communi-cation and social reciprocity, as well as indicates the patriarchal domination of women, and is thus an affront to universal female dignity. Feminists around the world can therefore happily promote a universal ban, and any counterargument is rubbished as an injustice that can be perpetrated under the guise of cultural tradition.[3]

---

[3] Thus, Susan Moller Okin (1999) points out that cultural defences can violate women's and chil-dren's rights and that there is a deep tension between feminism and multiculturalism.

Yet Nussbaum's view is that the real reason for a ban is fear of the Muslim 'other'. Bringing forward very Western democratic arguments around communication and feminism suggests a certain cultural attempt to impose restriction – i.e. to control that fear – under the guise of promoting a universal value, that of the dignity of women. The *burqa* is a useful example of this kind of clash between what Westerners see as a universal rule – the right of women to wear what they want – and a relative rule, such as that a modest Muslim woman may *choose* to wear the *burqa*. The important word here is, of course, choice. Ignoring this, the universalist implies that no woman would want to wear the *burqa* (a patronising attitude for feminists and others to take).

The majority French ban on the *burqa* is a highly relativist position, not a universalist one about dignity. The issue is not one of cultural relativism, but questions how far individual choice can be extended. If individual choice differs markedly from a universal rule, must it then submit to that universal rule (Lehrer 2001)? The question can be phrased more simply: how can the universalist ensure she or he isn't simply applying a conformist notion which may imply an alleged majority bias?

This leaves the global ethicist looking for some form of synthetic or mid-point theory that develops the negative tension between universalism and relativism into a more positive philosophy.

Communitarianism and cosmopolitanism are often promoted as likely solutions.

7.2 Communitarianism and Cosmopolitanism

At the risk of drowning in a sea of 'isms', one might start by noting a confusion of terms – relativism, pluralism, communitarianism, and particularism are often and erroneously regarded as the same (see Table 7.3 for the general usage of the terms). The term *cosmopolitanism* is the more useful one for the purposes of this discussion (Table 7.1).

Communitarianism is useful for the purposes of this study, in that it includes an idea of multi- or dual identity, as developed to a large degree by Charles Taylor, but also informed by the works of Habermas, Benhabib, Korsgaard, and Beck, among others.

The notion of multi- or dual identity also appears in the theory of cosmopolitanism, which is often promoted as a form of 'soft universalism'. Seen as a less hegemonic and monologic form of universalism, it is based on a balance of the universal and the relative, of community and alterity, for the following reasons:

1. We are world citizens today as well as national individuals with specific cultures (i.e. we have a dual identity).
2. Globalisation is a dual process, or dialogic, as will be explained.

Diogenes, when he referred to himself as a cosmopolitan, did not know that he might be spawning a global movement of 'world citizens', and indeed Diogenes himself left no blueprint for such citizenship. However, Kant's notion of

Table 7.1 A 'sea of isms': definitions

Universalism	The theory that there are universal values such as 'it is always wrong to kill another human' that can (and should) be applied across cultures
Relativism	The theory that there are cultural influences that may be taken into account, for example, some countries apply the death penalty
Pluralism	That both of the above ideas can be equally respected within society
Particularism	Primary allegiance is to the values of a group such as one's family, friends, or neighbours
Communitarianism	The socially embedded self requires individual views such as a belief in euthanasia to be balanced with potentially opposed views of the majority: 'Communitarians begin by positing a need to experience our lives as bound up with the good of the communities out of which our identity has been constituted' (Stanford Encyclopedia of Philosophy 2001). Sources of selfhood are taken from a range of social constructions and experiences
Cosmopolitanism	A dialogic theory that looks to mediate between universalism and relativism and between being a 'global citizen' and being a culturally defined individual. There are several types – this study uses contemporary definitions of moderate moral and cultural cosmopolitanism and particularly that of Appiah (2006) who sees cosmopolitanism as both a notion of shared citizenship and respect for individual beliefs

cosmopolitan law suggests that in addition to one's national citizenship, we also have citizenship of the world state, which operates on the basis of international law (Pojman 2005).

However, cosmopolitanism today has a more general meaning, one that has been called utopian in the negative sense of being ineffectively broad. It can mean global free trade and/or a single global market, as well as a rejection of exclusive attachments to particularist cultures, and/or a facet of modernity or postmodernity. Cosmopolitan ethics can also refer to either a sociological ethics (e.g. Beck or Miller) centred on the notions of citizenship and global solidarity, a cultural trend (e.g. Appiah), or a political concern with global justice (e.g. Habermas, Rawls, Sen, and Pogge).

Beck differentiates between cosmopolitanism as a form of 'mundane interaction' that is a 'by-product of global institutions and agencies plugging into one another' and cosmopolitan *realpolitik* (Maharaj 2010; Miller 1998). He does this by looking at dialecticism, mentioned in the previous chapter, as a central methodological issue in terms of how the East and the West might cooperate. This is an issue that interests Beck and Sznaider (2006), who promote a 'real cosmopolitanism' that 'seek(s) to overcome dualisms by proceeding on a logic of "both-and" instead of "either-or"'.

In other words, such practical cosmopolitanism rejects both so-called Chinese uniformity and Western either/or dissensus for a both/and form that implies relativism *and* universalism.

Cosmopolitanism, also called 'globalisation from within', based on 'the transnationality that is arising inside nation-states', is seen by Beck and Sznaider as an answer to the problematic nature of universalism. In their view the latter is faulty in that it obliges us 'to respect others as equals as a matter of principle', yet without involving 'any requirement that would arouse curiosity or respect for what makes others different. On the contrary, the particularity of others is sacrificed to an assumption of universal equality'. Thus, Beck and Sznaider (2006) suggest a form of cosmopolitanism:

> ...conceived, elaborated and practiced not in an exclusive manner but in an *inclusive* relation to universalism, contextualism, nationalism, transnationalism, etc. It is this particular combination of semantic elements which the cosmopolitan outlook shares with the universalistic, relativistic and national outlooks and which at the same time distinguishes it from these other approaches.

This complex statement with all its 'isms' can be boiled down to the suggestion that universalism attempts to find some form of synthesis with relativism. It is akin to the argument of Appiah (2006), who also noted that cosmopolitanism, as such a both/and form, has unsurprisingly been criticised for being:

> ...all things to all people. Is this the secret of its resurgence: an infinitely flexible mission statement for politics in the global age: a badge of identity politics, deference to difference and high value placed on plurality? Should it be understood as a political philosophy, an individual orientation to the world, a societal condition, a political project, or a pejorative designation?[4]

Yet, Appiah (2006) has also said that 'we cosmopolitans believe in universal truth, too, although we are less certain that we have it all already. It is not scepticism about the very idea of truth that guides us; it is realism about how hard the truth is to find'. (Admittedly, this doesn't really solve the problem, but only suggests the willingness to try and do so.)

All-inclusiveness is not the only criticism of cosmopolitanism. Zolo (1997), for example, argues that cosmopolitanism does not 'go much beyond the optimistic expectation of affluent Westerners to be able to feel and be universally recognised as citizens of the world' and also underestimates 'the way in which Westernization is cultural homogenization without integration'.

As a more pluralistic version of universalism, cosmopolitanism can still be understood as less judgmental (Morin 2009). This is in line with the eighteenth-century associations the term held, when 'cosmopolitan' indicated:

> ...An attitude of open-mindedness and impartiality. A cosmopolitan was someone who was not subservient to a particular religious or political authority, someone who was not biased by particular loyalties or cultural prejudice. Furthermore, the term was sometimes used to indicate a person who led an urbane life-style, or who was fond of traveling, cherished a network of international contacts, or felt at home everywhere (Stanford Encyclopedia of Philosophy 2002).

Another criticism has been that cosmopolitanism is difficult to assess in action – that it is better in theory than in practice. Benhabib (2006) has attempted to answer this issue by suggesting that a process of iteration is required:

---

[4] See also Rumford (2007).

when citizens become convinced of the independent validity of cosmopolitan norms, they can *'reiterate'* these principles and incorporate them into democratic will-formation processes through argument, contestation, revision and rejection... My answer to the question as to how to reconcile cosmopolitanism with the unique legal, historical and cultural traditions and memories of a people is that we must respect, encourage and initiate multiple processes of democratic iteration.

'Determination and the norms of cosmopolitan justice' can be resolved through an internal process within nation-states, in terms of which cosmopolitan claims are iterated by the people until they become cultural norms and gain the standing of law by common consent. The nation or people show itself to be the *authors* of such laws rather than merely the subjects. This suggests, of course, a stress on agency and particularly on Habermas's theory of the legitimisation of laws through public discourse, as discussed previously.

Cosmopolitanism, therefore, has its detractors, but it involves the following positive notions:

- It is process-oriented.
- It aims to be practical.
- It aims at a methodology that works on a both/and dialectic of inclusivity, one that ignores neither relativism nor universalism but attempts to make them work together.
- It stresses agency (in Benhabib's version, at least).

Combining these positive aspects, we might argue that what is really needed for cosmopolitanism to work practically and globally is a process model of both/and, or dual, identity. Cosmopolitanism is a process of detachment from one's self (or from the relative position), not so that identity is swallowed up by universalism, but in order to achieve 'reattachment with multiple affiliations' (Cohen and Vertovec 2002). Seen in the contexts of Habermas's universalisation principle and moral identity formation, the idea of dual identity offers a way forward for global ethics.

7.2.1   Identity

One who knows himself and others will find out here that East and West are no longer separable (von Goethe 1972).[5]
...universalisation... can be understood as a process of putting together a collage of varying perspectives... 'Intersections of experience'... may lead to that universalising moment (De Castro 1999).

Alasdair MacIntyre (1984) proposed three questions that get at the heart of moral thinking: *Who am I? Who ought I to become? And how ought I to get there?* These

---

[5] *'Wer sich selbst und andere kennt wird auch hier erkennen; Orient und Okzident sind nich mehr zu trennen, Sinnig zwischen beiden Welten sich zu wiegen lass ich gelten: Also zwischen Ost-und Westen sich bewegen sei zum Besten'.*

might also form the core of discourse ethics, namely, how can one achieve a rational self-aware selfhood? Can one achieve a global and more 'ideal' selfhood? And how does one get there?

The problems of universalism and relativism can be seen as built into the self, which (as communitarians and particularists both argue) is a self conditioned by its culture and context; thus, it is irrevocably relative and renders universalism impossible. Yet, at the same time that emotional identity and attachments to families and countries are formed, there is the possibility of a universal and transcendent approach to global citizenship, a transcendental identity. But how?

Here one might look at pTA and at Habermas, particularly his 'universalising' principle of discourse between individuals.

The first point to recall is that, as the species ethic implies, humankind in Habermas's view enters into dialogue with a tendency towards consensus. Habermas sees the ability to coordinate actions through communication (language) oriented towards reaching agreement as the most fundamental characteristic of human beings as a species. The focus of discourse ethics is what is 'equally good for all':

> ...this 'moral point of view' constitutes a sharp but narrow spotlight, which selects from the mass of evaluative questions those action-related conflicts that can be resolved with reference to a generalizable interest (Habermas 1992b).

Recognising the equal rights of participants in the moral discourse is a principle of universal moral respect and one of egalitarian reciprocity (Benhabib 1993). Thus, every person who accepts the process of practical discourse also implicitly presupposes as valid the *principle* of universalisation itself – in other words, by accepting the notion of argument, one also accepts every rational being's participation in the process.

This identifies an *agency* quality or virtue of orientation towards agreement as universal.

7.3 Habermas's Universalisation Principle

> Reaching understanding is the inherent telos of human speech (Habermas 2004).

What exactly does the universalisation principle imply? Habermas's dialogue model has two components, which he labelled 'D' and 'U'.

'D' is defined by Habermas as the principle of reaching agreement through practical discourse. Universalisation ('U') occurs when:

> All affected can accept the consequences and the side effects that [the norm's] general observance can be anticipated to have for the satisfaction of everyone's interests, and the consequences are preferred to those of known alternative possibilities for regulation (Habermas 1991).

Stated yet another way, Habermas claims that something can be valid when the foreseeable 'consequences and side-effects of its general observance for the interests and value-orientations of each individual could be jointly accepted by all concerned without coercion' (Habermas 1991).

Thus 'only those action norms are valid to which all possibly affected persons could agree as participants in rational discourse' (Habermas 1998a).

Through the 'D' and 'U' principles, discourse ethics aims to be a consensual ethics. Habermas develops the (Kantian) idea that the universal 'ought' – of norms as ensuring equality of treatment, of being publicly defensible, that valid norms deserve recognition – by arguing for a principle that constrains all concerned to adopt the perspectives of everyone else in the balancing of interest.

Asking whether we can find truth or reach agreement, Habermas suggests that 'D' settles the first question, while 'U' (universalisation) aims to settle the second.

Does this mean that the universalisation principle is a function of language, in that it is part of that 'telos of human speech' that Habermas mentions? It has been argued that Habermas follows Pinker's development of Chomsky's concept of universal language. This is the argument that human thought takes place in a prelinguistic form of cognition that is the same for all human beings – thus, we all think in fundamentally the same ways, as the signified to which we attach the signifiers of language are universal (Pinker 1994). This seems a little simplistic, given the culturally loaded associations some signifiers can have, and Pinker does suggest that complex grammar is universal 'within a society' rather than universally. The comments that can be made on this idea are:

(a) That Habermas's views on universal language are by no means clear
(b) That this brings us to the ideas of neurolinguistics, a subject which is beyond the scope of this study[6]
(c) That Habermas examines the mechanisms of speech in terms of how validity claims can be substantiated and rendered acceptable to interlocutors rather than simply claiming universal referentiality

Such a claim as raised in point (c) is arguably missing Habermas's point somewhat. 'U' regulates argumentation among many participants; Habermas argues along the lines suggested by Apel that every argumentation rests on pragmatic suppositions from whose propositional content the principle of universalisation ('U') can be derived.

What this all means is that Habermas wants his discourse agents to undergo the process of understanding the other through debate, not through any universal referent. Words that have value associations – such as the good life – are a focus for the testing of rational ideas in the attempt to show interlocutors that they need to challenge their views (and their cultural referents, instead of agreeing on the universality of these referents).

---

[6] A recent Hong Kong study allegedly managed to 'reprogram' subjects' cultural mindsets through linguistic and visual programming. Does this suggest that neural 'representations of the self and more generally the brain's ability to internalise diverse cultural perspectives – the basis of what we call cross-cultural empathy – are flexible and dynamic works in progress? See Rolbin and della Chiesa (2010), Begley (2010), and Ng et al. (2010).

Rather than relying on a universal grammar, interlocutors look towards what is termed 'ideal communication'. This form of justification, rooted in pragmatic agreement, but formulated to transcend individuals, has been called transcendental-pragmatic and/or universal-pragmatic discourse (Watt 1975).

Ideal communication relies on ideal conditions for argument – in other words, no restrictions on rational and honest debate (as discussed in Chap. 4, on noncoercive fora). As well as such rules, it also relies on the idea of the lifeworld ('Lebenswelt'), which Habermas suggests as the realm of social life. As Finlayson (2005) explains succinctly, this is a sphere that is subject to change, and has shifting boundaries:

> The shared meanings and understandings of the lifeworld form a unity, but not a totality… The contents of lifeworld are open to revision and change…In principle there is no reason why eventually every part of the lifeworld should not be revised or replaced.

Communication is rooted in a shared lifeworld (Apel 2008).[7] The lifeworld supports communicative action in that, as a shared context, it creates a platform for consensus. The lifeworld thus offers a place for change and for shifting boundaries that mean consensus can be negotiated.

In conclusion, the universalisation principle is the expression through rational discourse of the innate human capability to achieve consensus. It implies not only logistics for discourse (rules for 'ideal' speech) but also the possibility of mutual change in the lifeworlds of interlocutors, who in effect shift the boundaries of their social spheres or lifeworlds towards a world of agreement.

7.3.1 Deriving the Universal

How does this shifting of boundaries towards a new intersection of horizons actually work? The following discussion (with some additions to Habermas from other commentators such as Taylor, Korsgaard, and Beck) suggests a five-stage process of universalisation formulated around the following activities:

1. Decentralisation
2. Recognition
3. Accepting the responsibility of reciprocity
4. Negotiation for consensus
5. Identity creation

As this list suggests, the process aims to create the condition of 'dual identity', in terms of which the lifeworld of the interlocutor shifts towards a new situation that includes a part of the lifeworld of the other. Thus, the interlocutor, while remaining authentic in terms of his or her original lifeworld and identity expressed therein, has the option of a new, shared identity.

---

[7] See also Scherer and Patzer (2011).

7.3.1.1 Decentralisation

We need to create space for people to speak to, and listen to, one another (Tutu 2012).

Decentralisation has three applications in this discussion, in that it relates to:

- Deliberative democracy
- Decentralisation of place
- Decentralisation of self

The first issue has already been mentioned in Chap. 4 in terms of the current argument that the social and ethical issues raised by science and technology are no longer the responsibility of a 'relatively closed bureaucratic-professional-legal world of regulation', but have been relocated within the public arena (Jotterand 2006). So-called 'upstream engagement' occurs when the public is involved prior to significant research taking place and before attitudes towards any new technology are established.

Other meanings of decentralisation are probably less clear. One meaning derives from the idea that universalists vainly attempt to create a neutral space, one that is culturally unbounded or, as Solomon puts it, 'the result of us trying too hard to be "above" any particular society and culture' has the result that 'in the name of universalism, (we) find ourselves nowhere at all' (Solomon 2006). Yet as Westwood (2002) notes, we need a 'decentred social space', 'a space of neutrality for practice of the politics of recognition'.

As Chinese critics have observed, 'universalism' can often mean an imposition of Western values; hence, the space within which universalism is discussed should be neutral (Yang 2012). Thus, the decentralist context of intersubjective interaction to emphasise 'that only those elements of mutual interest to all participants should be considered as sufficiently neutral and relevant for discussion – so that no 'subject – whether an individual or a group' can 'declare itself the embodiment of reason' (Davies 2007).

Postmodern conceptions of space suggest a 'recovery of space as a social space', but a space located in 'deterritorialised' flows of communication. The public sphere is a network of discursive spaces in society (Delanty 2000).

The third context to the word decentralisation derives from identity. Cosmopolitanism, as already mentioned, implies 'a conception of the self as a situated being, a person whose values and scope of moral concern have been shaped by history, social context, and political boundaries' (Jordaan 2009). MacIntyre (1984) notes the way in which virtues that contribute to human flourishing are worked out in a particular community or tradition.

Decentralisation operates on the basis of recognising that 'sited-ness' within a community or tradition, yet also of attempting to place discourse with a wider, decentred or decentralised 'social space'. One might digress slightly to Jameson's interesting concept of 'mapping', which combines a sense of where one is with a sense of unbounded, utopian space. Arguing for the postmodern subject's inability to map him/herself spatially or politically, Jameson (1988) suggests that our awareness of situation may reveal those ideologically derived imaginary relations that conceal reality from the subject; thus, the 'mental map of city space . . . can be extrapolated to that mental map of the social and global totality we all carry around in our heads in variously garbled forms'. The notion itself is usefully dialectic and dialogic. Cosmopolitan identity implies a process firstly of detachment and then, secondly, a 'reality of (re)attachment, multiple attachment, or attachment at a distance' (Robbins 1998).

A decentralised space is one with a neutral viewpoint, a development of the idea promulgated through cosmopolitanism, which has been identified with the adoption of an impartial viewpoint; 'its crux is the idea that each person is equally a subject of moral concern, or alternatively, that in the justification of choices one must take the prospects of everyone affected equally into account' (Beitz 1994).[8]

Here one needs to be clear about the difference between 'viewpoint' and 'identity'. Wolfe (1992), for example, sticks to the issue of recognising viewpoints when she describes a member of the pacifist Amish sect who can appreciate the position of someone who thinks that violence is sometimes justified, while the other can appreciate the Amish position; each can acknowledge that the other's position is rational in its own terms, and yet each can regard himself/herself as justified in believing and following his or her own code. Wolfe (1992) suggests that persons can (sometimes) accept an outside perspective on their judgments without undermining their ethical convictions or succumbing to subjectivism. This defence depends on the assumption that conflicts are not total, i.e. that individuals will be able to agree despite differing opinions.

However, this study takes the line that viewpoint and identity are closely linked in discourse. This is not a particularly Habermasian view, as he falls back on logical argument to transcend personally held beliefs; yet it is a fact that such beliefs are strongly held, often due to being a part of who we are, and that they make rational argument and compromise difficult. I see the two as combined; thus, multiple identities can provide multiple viewpoints.

The postmodern self is a plurality, a protean 'multi-self', decentred' (Anderson 1997; Gergen 1990), or even 'extracentred, so that the relation of "I" and the other can be imagined as one of compossibility' (Venn 2002). While we are on this postmodern level, one might also mention Dower (2005), who notes the emergent transnational space of global civil society,[9] which in its dialecticism moves seamlessly to holistic dialogism, from the productive tensions of either/or to the fluid decentralisation of both/and.

Yet, in very simplistic terms, the 'decentralisation of self' means rising above one's culturally bound identity. The interlocutors in discourse/pTA need a critical distance from the normative assumptions of their particular culture. Hendley (2000) describes this as a process of 'abstraction', i.e. of abstracting the self from one's context:

---

[8] See also Scanlon (1982).

[9] Dower however depends rather on rationality and on liberalism in his arguments.

> Properly moral principles are formed therefore, only by a process that abstracts from the particularity of one's culture and its specific ethical ideals. Our concern for the concrete other, with a caring appreciation of his/her unique concerns and aspirations, necessarily disappears at this level of abstraction.

This may be more difficult than it sounds. Lukes (2008), looking at the alleged contrast between Western individualism and Eastern holistic modes of thought, examines the crucial issue of whether contrasting values and modes of thinking are 'culturally ingrained or can be switched off and on'. In other words, can the culturally bound(ed) self be transcended? Lukes argues that culturally ingrained values may not necessarily be deep or stable, making the point that there exists moral diversity in an individual's life, as well as the possibility of moral change or evolution – possibly towards universalism.

The dialogue structure, as Habermas predicates it, recognises reason, otherness, and self-awareness, but also suggests a mechanism through which universalisation might occur. As Habermas stated:

> Relativising one's own form of existence to the legitimate claims of other forms for life, according equal rights to aliens and others with all their idiosyncrasies and unintelligibility, not sticking doggedly to the universalisation of one's own identity, not marginalizing that which deviates from one's own identity, allowing the sphere of tolerance to become ceaselessly larger than it is today – all this is what moral universalism means today (White 1988).

Expanding 'one's sphere of tolerance' (which is not of course the same as respect, although Habermas (2002) implies that) does sound useful for attempts at consensus. But one still has a nagging feeling that in addressing a pTA group, more detail might be required on *how* exactly to do this. The better way to see this process of 'tolerance' is as suggested, one of 'decentralisation', of creating a neutral space in which no one 'tolerates' another, but in which interlocutors – as Habermas is at pains to stress elsewhere – meet as equals. Perhaps the 'tolerant person' should rather be understood in the sense that Benn (2001) uses the concept, as someone who is:

> usually not abrasive in the delivery of his judgments, is prepared to give others the benefit of the doubt, and is inclined to regard them as sincere and rational until shown otherwise. He pursues his disagreements by means of argument rather than force or abuse, and is open to the possibility that he may be wrong himself. But he does not thereby come to believe that his own views are no more true than the views of those who seem to disagree with him. It is possible to be tolerant while believing that others, to put it brutally, are simply in the wrong.

McCarthy's (2003) 'reflective equilibrium' modifies one's individual narrative, as MacIntyre suggests, to express a wider coherence values embedded in the communities within which the individual lives. Simply put, through reflective dialogue, we discover common ground. Taylor (1992) suggests that we must:

> learn to move in a broader horizon, within which what we have formally taken for granted as the background to valuation can be situated as one possibility alongside the different background of the formerly unfamiliar culture. The 'fusion of horizons' operates through our developing new vocabularies of comparison.

Zeiler (2009), attempting a 'theoretical underpinning for a bioethics that recognises the diversity of traditions and experiences without leading to relativism', suggests searching 'principally for sameness'.

This brings us to the issue of recognition.

7.3.1.2 Recognition

> What we should chiefly desire is to find ways to empower ourselves, individually and collectively, that also connect us, and ways to connect us that also empower us (Sen 1999).

The name associated with the 'politics of recognition' is often that of Charles Taylor, with his notion of 'other-understanding' leading to a 'fusion of horizons' – a phrase borrowed from Gadamer's 'Horizontverschmelzung'. Taylor uses it somewhat broadly, given that Gadamer's initial usage of the term is contextualised in a discussion of the intersection of history with the present, but the concept offers a succinct view of one rising above the self. Gadamer (2004) states that:

> Transposing ourselves consists neither in the empathy of one individual for another nor in subordinating another person to our own standards; rather, it always involves rising to a higher universality that overcomes not only our own particularity but also that of the other. The concept of 'horizon' suggests itself because it expresses the superior breadth of vision that the person who is trying to understand must have. To acquire a horizon means that one learns to look beyond what is close at hand – not in order to look away from it but to see it better.

Understanding happens when our present understanding or horizon is moved to a new understanding or horizon by an encounter; 'understanding is always the fusion of these horizons supposedly existing by themselves' (Gadamer 2004).

Before the phrase 'politics of recognition' derails the discussion, it can be noted that Taylor's work on recognition and multiculturalism is often used in political matters such as indigenous land rights; yet the idea of 'recognition' often means a sociological or ethical concept as opposed to a political one, and so this discussion uses it in those terms.

Gadamer's approach, of an I-Thou relationship characterised by openness and respectful questioning, hopefully allows one to recognise and respect the other as both the same and different, using a language based on a 'moral economy of interdependence' (Robertson). Let us examine the notion of *recognition*.

As Beck's theory of cosmopolitanism suggests, the willingness to work through dialogue rather than imposition needs to be matched by a willingness to move beyond narrow nationalism with its exclusion of 'otherness'. His definition of cosmopolitanism therefore implies the 'dialogic imagination, the capacity to explore creatively the contradictions within and between cultures' (Featherstone 2002).

Recognition, in terms of Gadamer's I-Thou, or Beck's 'creative exploration of contradictions', obviously suggests a recognition of *difference*. This means overcoming that well-known instinctive distrust of the 'strange' (that in its extreme form manifests as xenophobia). This involves a genuine struggle to recognise the validity and reality of others' views, moving beyond the instinctive reaction of 'that's stupid' when faced with a different view of how something might be done or intellectualised (Turner 2002). Charles Taylor's (1992) ethos of 'other-understanding' suggests that to 'really' encounter the other – as opposed to encountering one's own view of the other – we must 'give due acknowledgment only to what is universally present – everybody has an identity – through recognising what is peculiar to each'; he suggests an active effort to recognise difference.

As Todorov (1993) argued when discussing Rousseau's 'good universalism', it is based on dialogue and on becoming 'thoroughly familiar with the particular' so that the universal, the 'horizon of understanding between two particulars', may therefore be achieved.

Recognition is only the first step; a particular kind of recognition is meant, one that encourages the cultivation of:

> ...an unconditional desire to view and harness other people's uniqueness and difference, not as a threat but as a complement to one's own humanity... The relation with the 'other' is one of subjective equality (Eze 2011).

Taylor (1995) saw the process of understanding the differences between world-views as one of 'letting people be' or of not asserting power over difference. Recognition of people for what they are allows us 'to reach a common language, common human understanding, which would allow both us and them undistortively to be'. But here one might argue with Taylor. Recognition is not meant to be an appreciation of everything about the other, as Taylor implies, but the recognition of 'humanity'. In other words, it is a search for recognition of the principle of humanity. This implies, to borrow a phrase from Rumelili (2008), that we need to distinguish between 'good' and 'bad' 'forms of othering'.

One might conceptualise the idea of 'bad othering' as an utterly bland, passive acceptance of the other compared to 'good othering', which is an evaluative recognition, an active attempt to understand and to relate.

Rumeili discusses identity formation through differentiation (recognising the other), and here we come to the positive side of the process – once one has gone to all the effort of appreciating another's humanity, the payoff comes in the form of a greater sense of self. As Taylor (1995) rather more obliquely puts it, 'in the case of the politics of difference, we might say that a universal potential is at its basis, namely, the potential for forming and defining one's own identity'. Mead's theory of identity formation (from which Habermas was later to borrow) sees this as taking on the role of community identity. In his work on the genesis of the self, Mead argues that taking the role of the other is the mechanism that allows the development of self-consciousness, in that taking on the attitudes of the community means constituting the self from these same attitudes (Aboulafia 2001).[10] Finally, Merleau-Ponty also has a definition of this process. In Madison's (1997) view, Merleau-Ponty's ethics are founded on the notion of the self and other as coexisting symbiotically, 'reciprocally confirming each other in their own being' through an ethics of 'revers-

---

[10] See Mitchell Aboulafia (2001) for a discussion of Habermas and Mead.

ibility or reciprocity', in terms of which the self cannot be known until recognised by the other:

> In the experience of dialogue, there is constituted between the other person and myself a common ground; my thought and his are interwoven into a single fabric, my words and those of my interlocutor are called forth by the state of the discussion, and they are inserted into a shared operation of which neither is the creator...We are collaborators in a consummate reciprocity. Our perspectives merge into another [cf. Gadamer's notion of a "fusion of horizons'] (Merleau-Ponty 1962).

Taylor (1992) thinks identity is created dialogically, through our interactions with others: 'If human identity is dialogically created and constructed, then there is room for us to deliberate about those aspects of identity that we potentially share'. Recognition thus implies not only an active process of understanding, tolerating, and appreciating the other, but recognising that reciprocal humanity. Recognition of difference is accompanied by recognition of sameness.

Before looking at reciprocity, however, one should note that the recognition process has been criticised as unfeasible.

It has been argued that Habermas's (1993) interest in cultural specificity, the lifeworld of the interlocutor, meaning that he/she is embedded in the argumentative praxis of a specific social world with a specific cultural and historical legacy, renders the possibility of later universalisation unlikely.

Equally, critics condemn Benhabib, as her development of Habermas's ideas still does not demonstrate how recognition of specificity develops into universalism:

> Habermas...situates rather than universalizes the conditions for truth... Benhabib's attempt at overcoming the same dualism by demanding a focus on the other has similar strengths and weaknesses...(as) the process of recognising the other is not natural and automatic, but depends on socially variable conditions. Thus Benhabib merely displaces the problem of universalism on to these new procedures for judgment which are not sufficiently universal to be adequate to the task demanded of them. The 'act of recognition' requires a social process of assessment as to what constitutes the same or different from oneself (Walby 2002).

Linking moral action to its cultural basis implies a reduced scope as the focus shifts towards local problem solving (Schneider 1994). In short, we can never transcend our own historical and cultural situations.

Yet Habermas (1998b) defended what he calls a 'correctly understood universalism' by noting that universalism need not assimilate heterogeneity by levelling cultural differences, since a 'correctly understood community' constitutes itself through the 'negative idea of abolishing discrimination'; 'inclusion of the other' implies community openness as well as free and equal recognition.

Perhaps a simple way of characterising recognition is through the idea of hospitality, which has been suggested by some theorists as the way forward for cosmopolitan ethics (Turner 2002). Derrida's (2001) cosmopolitanism (although perhaps unhappily Eurocentric) foregrounds the notion of hospitality in the ethical welcoming of the stranger and is a neat, practical way (within certain parameters of the word 'practical') to ground our recognition of otherness.

As Habermas might put it, the universalisation principle requires participants to attend to the values and interests of each person as a unique individual:

conversely, each individual conditions her judgment about the moral import of her values and interests on what all participants can freely accept. Consequently, moral discourse is structured in a way that links moral validity with solidaristic concern for both the concrete individual and the morally formative communities on which her identity depends (Stanford Encyclopedia of Philosophy 2007).

Webb (2006) might refer to the 'longstanding European tradition of humanism with its various considerations of shared moral understanding, the cultivation of the self, mutual reciprocity, social virtues, and the common good' in his discussion of virtuous practice, but in fact, the notion is not humanism as much as recognition of humanity. It is an active process, in the sense that it implies, as Eze has said, a desire to utilise the human potential recognised in the other, presumably aimed towards mutual flourishing.

7.3.1.3   The Responsibility of Reciprocity

Habermas situates rational collective will outside formal organisations, for discourses do not govern. They generate a communicative power that cannot take the place of administration but can only influence it (Calhoun 1992).

Mutual flourishing requires a sense of responsibility, or obligation, towards the other as he/she flourishes alongside me. This is a tenet of much Western philosophy, in terms of Christian responsibility for others, but is also found in Confucian ethics. Confucius was alleged to have responded to the question of whether there was 'any single word that could guide one's entire life' with the word 'reciprocity' – 'what you do not wish for yourself, do not do to others' (Yu 2003). As Madison (2002) notes, this view is a 'prototype of the ethics of mutual recognition called for today by the logic of globalization'.

However, reciprocity suggests a rather idealistic view of humankind, given that multiple examples from daily life can demonstrate how individuals have exploitative desires focused purely on their own good, often at the expense of other people's flourishing. It implies that all human beings possess some form of basic virtue, their humanity equating with a concern for others. Jonas (1984) might have a better way of looking at this. In his 'Verantwortungsethik' he suggests that to feel responsibility, one must become aware of the fragility of the other, i.e. that we hear the call of the other's 'perishability, indigence, and insecurity'. This requires a moment of care and apprehension for humankind, prompted by fear. Similarly, Rorty's (1989) view of human solidarity promotes the idea of 'increasing our sensitivity to the particular details of the pain and humiliations of other, unfamiliar sorts of people'.[11]

Jonas's 'heuristics of fear' suggests that we share the fear of our mutual human vulnerability – which is thus the source of responsibility. It can be argued though

---

[11] See also Simpson and Williams (1999).

that it is precisely that sort of fear that one would rather not acknowledge and that denial is very much a twenty-first century mode.

Habermas suggests the basic virtue is a willingness to communicate, with the intent to reach agreement – a virtue that implies openness and receptivity. This does not necessarily imply responsibility and obligation, except in a limited sense, such as that of participating in dialogue. Cortina (2008) similarly states that an awareness of one's responsibility to participate in public debate becomes the crucial point for a 'phenomenisation' of civic morality that brings citizens together.[12]

If we look back to cosmopolitanism for some help on this matter, as Beck and Grande (2010) have noted, it is insufficient merely to *want* 'good'; there must also be a cosmopolitan ethics of responsibility. The theories of Weisband (2007), who develops discourse ethics with an implied notion of responsibility, provide a more overt idea of discourse accountability.

Weisband, having stated that he believes in a decentralised approach to ethics, or, as he describes it, a networked approach that encourages participation, is less hierarchical and recognises 'nebulous or uncertain boundaries of governance'. He then suggests something termed 'reciprocal intersubjectivity', discussing how the self and other are 'reciprocally exchanged' through virtue and 'reversibility' (or 'becoming' the other'), for:

> it is through the language of reciprocated accountabilities that we construct our social or collective identities in ways connected to the progression of a postmodern public ethics. And this becomes a key to our display of virtue . . . through accountability practices and relationships (Weisband 2007).

Weisband's ideas, despite their merits, are limited in application. Offering accountability practices as 'a kind of postmodern text written about appearances', Weisband (2007) mentions the technique of naming and shaming as one useful accountability-promoting technique, given the effectiveness of credibility loss as a sanctioning mechanism in global postmodern societies. This may be partially effective against corporations, depending on how much public outcry affects their market share, but less so on the many other global issues that require agreement.

Gutmann and Thompson (1996) distinguish between reciprocity and accountability, seeing the first as expressive of a sense of mutuality that citizens bring to a public forum and the second as a part of democratic deliberation (individual responsibility).

Korsgaard has a different approach, linking agency to identity via morality. To explain, obligations arise out of the practical identities we maintain in our everyday life – in other words:

> When you deliberate, it is as if there were something over and above all your desires, something which is you, and which chooses which desires to work on. This means that the principle or law by which you determine your action is one that you regard as being expressive of yourself (Korsgaard 1996).

---

[12] See also Apel (2008), which looks at the need for a universal ethics as a call for a mobilization of moral responsibility on a global scale, agreement on which can be achieved through discourse ethics.

Thus, to be oneself is to be a moral self. In fact, any 'practical conceptions of your identity which are fundamentally inconsistent with the value of humanity must be given up'; she gives the example of an assassin, whose identity does not derive from the value of humanity he demonstrates. Korsgaard adds a further dimension by arguing that one's moral identity is practically arrived at through social interaction. One perceives obligation as necessary due to threats to one's identity – in other words, ignoring social obligation means accepting such a threat to yourself:

> The conception of one's identity here is not a theoretical one . . . It is better understood as a description under which you value yourself, a description under which you find your life to be worth living and your actions to be worth undertaking. So I will call this a conception of your practical identity (Korsgaard 1996).

This is all, however, a little too vague, just as Habermas's species ethic was in describing how we innately desire consensus. Dostoevsky, would have reminded such philosophers that the human being can be innately perverse, desiring neither consensus nor a practical moral identity. Korsgaard et al. (2009) argues though that the 'process of self-constitution' requires us to act morally. A commitment to the moral law is built right into the activity that by virtue of being human, we are necessarily engaged in the activity of making something of ourselves. The moral law is the law of self-constitution.

Given that responsible reciprocity is not clearly defined, let us look then at self-creation as another and perhaps more useful stage in this process of 'deriving the universal'.

Before doing so, the notion of compromise (as the outcome of) requires brief discussion to see whether/if that contributes to identity creation.

7.3.1.4   Negotiation and Consensus

> Incomprehension and misunderstanding, intentional and involuntary untruthfulness, concealed and open discord can distort dialogue and prevent consensus (Habermas 1979).

The process of examining cultural differences, achieving dialogue, and then attaining compromise sounds fairly easy. Or is it? To understand oneself is a process of unmasking negotiation temptations to 'to deny our mutuality', of learning that although 'you and I are irreducibly different from each other', as human beings, we are of the same basic constitution:

> Our common constitution demands mutual recognition. Nonetheless, because our vulnerabilities are never eliminated, we must constantly struggle to achieve it. This is a struggle against the misrecognition of others at the same time that it is a struggle for recognition of oneself by others (Ricoeur 2005).[13]

---

[13] See also Ricoeur's (1992) *Oneself as Another*.

Presumably, in Habermasian terms, such vulnerabilities are counteracted by rational discussion. Let us examine what Habermas has to say about reasoned debate aimed at consensus.

Habermas suggests that relevant reasons should be acceptable to any reasonable agent. This takes the notion of one's subjective reasons a little further than the sphere of testing through assessment of sincerity, onto a more universal level instead. Habermas presupposes that a universal (ideal) consensus results if participants argue reasonably (and presumably for long enough so that all arguments are tested).

He does, however, appear to admit that ethical claims may be difficult for consensus, for example, because many are bound to introduce questions of the good life for individuals or groups and are therefore value based and so situationally determined (in that an individual's view of the good may be largely determined by his/her education and upbringing, class structure, nationality, and so forth).

It can be noted, however, that one class of ethical questions for which Habermas (2003) requires universal consensus are species-wide ethical issues relating to the nature of the human being raised by new technologies such as genetic engineering.

However, where different discourses lead to competing conclusions, or when issues arise in which discourse becomes unclear and/or deeply contested, Habermas appears to rely on the basic notion of his discourse theory – that we communicate in order to understand and reach consensus, rather than to fight for deeply held beliefs.[14] Where disagreement arises, this is a 'distortion' of discourse, meaning that there has been a derailment of the process, perhaps a lack of understanding and a lack of sincerity or validity. Habermas (1992a, b) insists on the universality of the grammatical role of concepts such as truth, rationality, or justification, regardless of their differing interpretations.[15]

This may seem rather idealistic, for dialogue leading to consensus is a process of disagreement. Consensus might never be reached. Yet, as Gutmann (1993) argues, though there may be irresolvable cultural differences about some moral matters, the more significant issues (at least as far as political ethics, which is her topic, is concerned) are the fundamental matters on which individuals agree:

> No culture or political community with which we are familiar gives its members good reasons for rejecting principles or practices that protect innocent people from being enslaved, tortured, murdered and malnourished, imprisoned, rendered homeless or subject to abnormal physical pain and sickness.

As Hutchings (2010) argues, any approach to global ethics must start with the problem of 'identifying and achieving justice in a context in which right-minded people disagree', though 'the question of what is right in some cases lacks a unique and determinate answer' (DeCew 1990).

---

[14] See McCarthy (1978, 1991) for commentary on Habermas and the issue of contested arguments.

[15] See also Habermas (1993).

The principles already discussed, those of decentralisation, negotiation (forcing opponents into decentring their perspectives by recognising the validity of other perspectives), and reciprocity, suggest that there are forces that counteract 'unresolvable' argument. There is also, arguably, room for answers rather than one answer; we are reminded of Lyotard's postmodern conception of justice as distinguished by divergence, multiplicity, contestation, novelty, and opinion of an ethical theory the task of which is 'to recover principles from tradition and to clarify the meaning of local norms and institutions' (Fairfield 1999).[16] As Habermas notes, compromise may be a dynamic and constantly evolving process rather than a final achievement.

Gutmann and Thompson (1996) also suggest a partial consensus, arguing that participants in democratic discourse are not being asked to change their first-order beliefs, but:

> To discover what aspects of those beliefs could be accepted as principles and policies by other citizens with whom they fundamentally disagree. Since it is this second-order agreement that citizens should seek, they do not have to trade off their personal moral views against public values.

Bargaining, or working out disagreements with a view to individual interests, is common in deliberation. Gutmann and Thompson argue for its value, as long as it is constrained by the overall value of reciprocity.

Dialogue results in openness to the other, 'recognizing that I myself must accept something against me' in attempting to fuse horizons (Gadamer 2004). This learning process can be characterised by the example of the three-stage model of communicative cultural integration proposed by Wohlrapp. In the first step, actors have to submit themselves to the 'experience' of the other as something beyond the familiar. In the second step, the actors strive to achieve an 'understanding' of the experienced unfamiliarities, and a third step aims to 'produce peaceability' (Wohlrapp 1998).

Dialogue ethics innately suggests compromise. Habermas (1991) defines consensus in a way that suggests that it may not be ideal, but merely 'preferred', a state wherein:

> All affected can accept the consequences and the side effects its general observance can be anticipated to have for the satisfaction of everyone's interests (and these consequences are preferred to those of known alternative possibilities for regulation).

Another way of looking at this issue of 'open' consensus is *dialectically*. This term is used in a broad sense in this study to mean a creative tension between opposites (including between the notions of agreement and disagreement). Dialecticism is based on the idea of dialogue and the dialogic/dialectical self:

> ...the self is not merely a recipient of cultural forms, passively absorbing a range of values... Agents participate in a cultural world, engaging in transactions of meaning and furthering the social reproduction of various ways of understanding the world... The core

---

[16] Georgia Warnke (1993) discusses 'hermeneutic conversation', the 'idea that an interpretive pluralism can be educational for all the parties involved'.

claim in the dialogical conception of the self is that the self is a product of social interaction (Wisnewski 2008).

Out of such dialogue and dialogic activity, a dialectic can be developed. Thus, Siep's fluid dialogue between universal and local, Gill's (2005) Stoical/Aristotelian model of a 'complex, forward-and-back negotiation between localized and universal norms' in his pluralist approach to the problem, Etzioni's productive tension between individual and community, and Beck's 'glocalism', all suggest a harmonious relationship between consensus/dissensus. (Multiculturalism, in Beck's (2002) view, asserts plurality, fostering a 'collective image of humanity in which the individual remains dependent on his cultural sphere'; being local yet cosmopolitan, he is 'glocal'.)

Consensus derives from the decentralised space that is created from such fluid dialecticism.

Another way of looking at fluid dialecticism is to examine the concept of negotiation, which, as Gutmann and Thompson have already argued, can operate as a second-order process that may be in a dialectic relationship with first-order beliefs.

There are two ways in which negotiation might be used as a term in this discussion: as negotiated identity and as a negotiation strategy. The point, however, is that in terms of discourse identity, the two are closely linked.

Dhanda, arguing that identity is negotiated, sees it as a combination of Rorty's flexible bargaining method of developing identity, combined with Taylor's view that there are underlying 'moral ontologies' that reveal our sense of the truth of who we are. Dhanda uses the useful analogy of a stranger in a town – the 'map' he might use to find his way correlates to Taylor's belief that there is a background situation which is fixed, whereas when trying to find a specific building, you might supplement the map by asking directions, taking advice, and seeking other's guidance while attempting to find the new address. This is another version of Gutmann and Thompson's double-order process of deliberation.

Thus, Dhanda (2008) comes up with the 'givens' for a negotiation strategy, of which the most relevant is that there must be an acceptance of 'fixed' identity as well as open identity, for 'our practical identities locate us in a particular time and place... all our negotiations for identity must proceed from this location'.

There is also a pragmatic element, explained best perhaps by Sen and Williams (1982) when they note that we share a social space and that such sharing gives us:

> ...some shared understanding of the psychological bases of moral agreement and disagreement themselves; a sense of the virtues, of expected conduct, or of public principle, and with these we work, in seeking to articulate and perhaps resolve disagreements.

The idea of sharing suggests that we may look at the idea of shared versus individual identity.

7.3.1.5 Creating Identity

> The self... is a process of socialization that itself already presupposes the structure of relations of reciprocal recognition (Habermas 1993).

Sen's (1987) attempt to create a new approach to the issue of global justice, one that seeks a way between the established approaches of 'grand universalism' and 'national particularism', in terms of so-called plural affiliation, predicates identity as a nexus of global and national aspects of affiliation (Robeyns 2002). Appiah's cosmopolitan approach requires people 'to be capable of living double lives, of stepping outside themselves, of taking a position both in a culture and external to it' (Neem 2009). More specifically, Beck argues that cosmopolitanism requires an epistemological shift, for we need to learn how to operate in two frames of reference: the local and the global (Werbner 1999). To do this, we need to cultivate a different self, what one might call a 'global' or decentralised self, less firmly tethered to time and place. The self is defined in (post)modernity as radically undogmatic, aware of limitations, and dual, i.e. both a universal self and a situated self.

This can be seen as adopting, in addition to one's culturally bounded self, the identity of a 'global citizen'. It also raises the issue of global moral duty, as discussed by Walzer, O'Neill (2001), Singer (1972), and Nussbaum (1996), for example, in terms of whether global moral obligations are as important as one's national or local duties. In other words, has this idea returned us most unhelpfully to the universalism/relativism debate? Miller (2002), in suggesting that 'international law must overcome its apparent bias in favor of states and open its institutions, procedures, and principles to individuals who might then act as "world citizens" by accepting moral responsibilities towards humankind as such', argues that:

> talk about the 'world community' can best be understood as an attempt to apply "communitarian", i.e., particularist principles of responsibility and obligations, to the universe of humankind. According to our conventional philosophical wisdom, this is an evident paradox. International solidarity would mean that states should assume the same kind and degree of responsibility for the wellbeing of any part of humankind as they reserve for their own nationals.

This does sound a little too idealistic, though.

Habermas's notion of a cosmopolitan public sphere suggests a post-national citizenship that is not anchored in territory or the cultural heritage of institutions, but involves an identification with the normative principles of the constitution. However, this gives it a political aspect, in that a democratic constitution is essential. On global citizenship, he argues a little vaguely that 'only a democratic citizenship that does not close itself off in a particularistic fashion can pave the way for a world citizenship, which is already taking shape today in worldwide political communication' (Habermas 1998a, b). In other words, the global citizen is simply a citizen of global discourse.

Rather, we are looking for a global identity that can be adopted in addition to one's first-order or culturally bound self. This is a version of Delanty's (2000) argument that:

> The decoupling of citizenship and nationality is strikingly evident in the question of identity…with identity increasingly becoming the basis of participatory politics…the politics of identity is more than just the assertion of identity…identity is not an inherent personality of cultural condition but something that can be freely chosen and is not exclusive. This emphasis on multiple identities is one of the main changes in identity formation today.

Not even multiple identities, but a dialogic or dialectic self, as not only does society create self, but the self in turn creates society. As Appiah (1996) notes:

> The self is…dialogically constituted because it is in dialogue with other peoples' understanding of who I am that I develop a conception of my identity… also because my identity is crucially constructed through concepts (and practices) made available to me by religion, society, school, and state, and mediated to varying degrees by the family.

Thus, the ideal discourse agent is someone with a dual identity that is dialectically fluid. How does one become this person? The principle of universalisation is intended to compel the *universal* exchange of roles that Mead called 'ideal role-taking' or 'universal discourse' (Mead and Morris 1934).[17] Habermas, it is known, took some of his ideas from Mead. In 'Individuation through Socialization: On George Herbert Mead's 'Theory of Subjectivity', (1992a, b), Habermas (1992a, b) extends Mead's idea of the socialised self whose identity is formed through relationships:

> … universalistic form of life, in which everyone can take up the perspective of everyone else and can count on reciprocal recognition by everybody, makes it possible for individuated beings to exist within a community – individualism as the flipside of universalism. Taking up a relationship to a projected form of society is what first makes it possible for me to take my own life history seriously as a principle of individuation – to regard it as it were the product of decisions for which I am responsible.

In other words, the individual must see himself/herself as responsible for his/her own morality, while socialisation, or recognition of the other, develops that process of (moral) self-recognition. He/she becomes someone with a dual identity that can be called 'intersubjective'. In other words, his/her dual identity is a reciprocal recognition of mutual, shared, or intersubjective humanity.

Therefore, the suggestion is that the universalism/relativism debate can be reframed in contemporary terms as part of a pluralist globality based on a decentred and dual identity, a 'reciprocating intersubjectivity'.

This can be expressed simply, in terms of values, as Yang (2012) does:

> The Confucian ethics code 'do not do to others what you do not want done to yourself' coincides with the…'Golden Rule' of the West… According to Confucius, the relationship between oneself and others is an interactive one, which is a form of inter-subjectivity, as expressed by him in 'wishing to be established oneself, one seeks also to establish others, wishing to enlarge oneself, one seeks also to enlarge others', and win-win is achieved through inter-subjective interaction.

Or it can be expressed as a methodological issue. Gan (2012) makes this point, arguing for the combination of universal principle and practical (i.e. method-based) adaptation to a situation as an 'effective combination' for 'people to handle and settle moral differences'.

In terms of methodology, what does 'dualism' mean? We have already discussed the idea of dialectic.

---

[17] See also Mead's *The Individual and the Social Self* (1982).

There is always a dialectic between self and society. Communitarianism, for example, while it emphasises the importance of community, is also very focused on the notion of otherness or difference: 'communitarians invoke Heidegger's *Differenz*, Derrida's *differance,* Lyotard's *differend,* Levinas's heteronomy, and Mikhail Bakhtin's heteroglossia', and advocate Foucault's 'opposition to any kind of universal rational ethics' (Lee 2001). In short, community implies pluralism, a group of selves and others. We find our group identity through our own identity, through identifying with, and noting our difference to, otherness; as Etzioni (2002) states, 'the dichotomous opposition between partiality and impartiality, or between particularistic and universal obligations, holds only if we assume that one's position on this matter must be all-encompassing', for 'in social reality people often combine the two orientations'.

Etzioni suggests that the particularistic is part of the communitarian, for particularistic obligations are part of the socially embedded self. And, although as Sandel (1998) argues, we understand ourselves as the 'particular persons we are – as members of this family or community or nation or people', communities are essential for one's particular identity. A particularist obligation includes the nurturing of a moral ecology within which communities function and within which moral ideas can be universalised. The relationship between individual and community, unless defective, will not necessarily manifest itself in dichotomy, but rather, as Etzioni (1996) suggests, through a 'productive tension'. Thus, we find our way to the development of an 'intercultural personhood' (Taylor 1992; Beck 2002). Kim (2008) explains thus:

> The term, intercultural identity, is employed as a counterpoint to, and as an extension of, cultural identity, and as a concept that represents the phenomenon of identity adaptation and transformation beyond the perimeters of the conventional, categorical conception of cultural identity…through prolonged and cumulative intercultural communication experiences, individuals around the world can, and do, undergo a gradual process of intercultural evolution.

For Habermas, moral development meant maturing beyond adherence to social contracts, to a new stage of guidance by universal ethics principles. This does not mean forgetting one's context or community-driven social contract, but rather that the developed human being is able to transcend the latter when it is considered unjust. Thus:

> rational autonomy supersedes societal concerns…at Stage 5, with its social contract …right consists of an awareness that people hold a variety of values and opinions… But for Kohlberg and Habermas, some non-relative values and rights…must be valued regardless of societal opinion. Stage 6 provides another example: The right choice is viewed as self-chosen, based on universal ethical principles (Cortsese 1990).

Is this theory of stage 6 based too much on Western principles of ego-development and autonomy?[18] Eastern critics have disagreed, stating that Kohlberg's 6 stages include 3 stages that are culturally universal and 3 that are culture-bound. In short,

---

[18] Kohlberg has of course been heavily criticised by feminist critics such as Carol Gilligan (1982).

Table 7.2 China/EU points of similarity/dissimilarity in nanotechnology discussions

East	West
Risk and economics are primary foci	Risk and economics are primary foci
	Precautionary principle applied
Economic issues in terms of consumer attitude	Economic issues but *also wider societal* issues, i.e. enhancement
pTA an *emerging context* for nanopolicy	pTA established for nanopolicy
Ren-ethics – benevolence	Principlism – i.e. do no harm
Approach to nano based on harmony or *tao*	Approach to nano based on reasoning and argumentation

the model itself is dialectical, involving *both* universalism and relativism, in a fluid structure that implies movement across the boundary between culture and universe and that can 'integrate' the two different perspectives of collectivism and individualism, while acknowledging their differences (Keung 1988) (Table 7.2).[19]

7.3.2 Identity Formation and Nanoethics

(It is necessary to)… shake off the yoke of national prejudices, to get to know men by their conformities and their differences, and to acquire that universal knowledge that is not exclusively of one century or of one country but of all times and of all places, and thus is, so to speak, the common science of the wise (Rousseau 1986).

How do all the above ideas fit within nanotechnology pTA?

Am (2011) looks at the German *NanoKommission* in terms of the creation of new trust relationships in the governance of nanotechnology, i.e. with a view to how a new 'we' in governing can arise among previously disparate positions transformed by 'new relations of trust and mutual responsibilities'. Am refers to a new ethos governing participatory governance, or democratic deliberative governance, suggesting that stakeholder fora like the *NanoKommission* can both offer a decentralised forum for responsible nanodebate and:

Contribute to the forming of new identities and the creation of a new style of governance that is marked by new relations of trust among actors who might previously have entered into controversies about an emerging technology…can contribute to the development of trust and mutual responsibility of the involved actors …bring about effects on the formation of boundaries of what is sayable and thinkable in nanotechnology governance (Am 2011).

Am draws from literature on identity building in deliberative policy and from Szerszynski (1999) on how trust transforms social identities and relationships. However, rather than suggesting how such identity transformations might work, Am moves to a focus on how the deliberative process becomes less transparent due to the concealment of political agendas within the forum. Thus, her article, while

---

[19] See also Dien (1983).

promising a way forward, also cautions against the way in which acting together in governance networks can produce restraints.

On decentralised or distributed responsibility, one aspect of the individual identity element of global ethics brings us to the individual responsibility of the scientist.

Kjolberg and Strand (2011) look at EU-CoC, The EU Code of Conduct for scientists (discussed in Chap. 3), as bringing 'the concept of responsible nanoresearch a long way' but as having one 'crucial element' lacking, namely, 'responsible nanoresearch as increased awareness of moral choices'. By this they mean that researchers' responsibilities might be assessed according to criteria that seem poorly defined.

And, as Subra (2011) notes, the issue is not simply one of research ethics, but also of disparate standards in reviewing such ethics: 'disparate standards for scientific merit review and differences in the infrastructure that ensure professional ethics and scientific integrity…are further exacerbated by cultural differences that arise from the large range of social perspectives and stages of national development'.

Von Schomberg's (2007) calls for 'new and badly-needed intermediate deliberative science policy structures', namely, forms of distributed responsibility that place more emphasis on the individual. The authors introduce Pellizzioni's (2004) notion of responsibility as a 'willingness to understand and confront the other's commitment and concern with our own, to look for a possible terrain of sharing', that 'entails readiness to rethink our own problem definition, goals, strategies, and identity'. Thus, while intuitional structures provide governance and direction, there is a need for nanoresearchers to question themselves and to act actively rather than reactively; moral education is therefore key for professionals.

It seems that educating the public and scientists on how to develop a dual identity might help with global nanoethics (Table 7.3).

The final word on this might usefully go to Eastern scholars. Habermas's views on moral education (which of course derive heavily from Kohlberg) have some similarities with Confucian teachings on moral self-development, from which Tran and Shen (1991) extrapolate the following Habermasian-Confucian stages:

- The learning process is one of mutual cognition and recognition.
- The learning process and cognitive reception are developed through individual as well as social praxis.
- Moral cognition or consciousness is acquired through learning fundamental human interests.
- Moral laws are constructed on moral judgment and the consensus of basic interests.
- The act of consensus is free from coercion.

As the Chinese philosopher Mencius noted, human nature is understood in terms of growth; Tang and Liang have argued that Mencius's philosophy involves dignity and criticism of one's society as aspects of the maturing process of the human being. This sounds rather similar to Western ideas of a morally developed society as dependent

Table 7.3 The five-stage process

Decentralisation and Recognition	Action(s): Global advisory body sets agenda. Local policymaking bodies agree to consider outcomes of forum. Agenda circulated to citizen's forum for commentary; inclusion of expert viewpoints; public survey to uncover most emotive issues as well as more mainstream concerns
	Methodology: Dialogue conducted via internet and other methods (for those without access to IT). Selection of agents (panel) to represent all groups, face-to-face if possible
	Intention: to uncover polarised viewpoints for debate/dialectic
Reciprocity, Negotiation, and Identity formation	All agents recognise the opposing viewpoint
	All agents asked to respond dialectically to the model, in terms of personal belief and general ('global') view, i.e. as 'I-self' and as 'we-self'
Thus	Dissemination of consensus by panel members to groups (constituencies) as well as dissemination of education strategy based on forum knowledge outcomes.

on ideas of self-worth and freedom in the realisation of one's own virtuous life and virtuous agency (Liang 2009). And as Fung notes, such agency is dual, given the 'dualism between the sanctity of personal liberty and the public morality of service to society and state'. He bases his argument (partly) on the work of the philosopher Hu Shi, who argued that individualism consists not only of a free and independent personality but also of self-development and responsibility – including social responsibility:

> Driven by civic virtue or public morality, rather than coerced by the state apparatus, the autonomous agent may see a higher value in collective interests than in private gains, in certain times and circumstances, taking serious actions to further those interests as part of a moral repertoire and as an expression of moral autonomy (Fung 2006).

Unbridled individualism is not useful; individualism that exists in a creative tension with greater concerns – what Hu called the immortal, greater self – is good. From the tension between the two comes an intercultural personhood that can, through dialogue, make global decisions and effect global change.

The combination of pTA and identity theory (intersubjectivity) suggests a model for nanotechnology dialogue aimed at consensus. This can work (within the established parameters for such a dialogue) across both East and West, for the Confucian virtuous agent is similar to the Aristotelian and the Habermasian in that both see moral virtues as derived not from personal discovery, but rather have an intersubjective view of ethics based on a dialectical relationship between self and others.

Thus, the pTA/Habermasian model of intersubjective ethics requires a dialectic between self as individual and self-as-other or self as collective.

Through the dialogical process of the pTA forum, a second self, or 'intercultural personhood', can be created, in terms of which the horizon of the nanoagent stretches beyond their lifeworld to encompass the horizon of the other.

It turns the usual negative of East-Western ethics from a negative (the individual and the collective are two very foreign notions) into a positive (the dialectic between individual and collective is what makes ethics work).

If we now return to nanotechnology, how might such a model apply? In short, why do I believe an enhanced pTA model might be useful for a global approach to nanotechnology policy?

Given that the emphasis on procedure, rather than values, is intended to make it easier for those with divergent views on nanotechnology's impact on, or contribution to, any 'good life', to achieve consensual progress, the question is first one of setting a practical agenda for a nanodialogue between East and West. As suggested on page 252, global nanodialogue agendas should be set by global advisory bodies. COMEST, as described in page 114, seems the most likely body (given that China is one of its members, as well as several EU countries). Given the neutrality of the prodecuralist model, which focuses on how debate is to be conducted rather than proscribing outcomes, the notion of 'agenda' would optimally become less a charged space of ideological confrontation and more a 'decentralised' arena for discussion.

Given the discussion in Chap. 2 of the difficulty of assessing the societal implications of nano, the main item on the agenda might be that of greater public education (rather than merely inclusion). In other words, the nano-question that might be proposed by global dialogue is: how are better-informed citizens created and brought into nanodialogue?

Here one must add that as well as being a practical question that can be procedurally debated using the 5-stage pTA model this thesis predicates, this question clearly exposes a 'clash of civilisations' or of Habermasian lifeworlds – thus such a dialogue is likely to be difficult before it even begins. Taking a broad view of East-West nanodebates thus far (as discussed, e.g. in terms of the lack of pTA to date in China), we could predict that 'better informed citizens' is a very Western notion unlikely to be fully embraced by all the Chinese participants in this dialogue (though it should be added that it might not be embraced by all the Western participants either, particularly given that scientists and policymakers included in this multi-voiced debate can be notoriously cautious of potential intellectual property theft.)

To make the 5-stage model work, the intersubjective identity-formation aspect of the 5-stage model comes into play.

Habermasian lifeworlds do not imply replacement of one by another, but of an intersection of horizons achieved through dialogue. The Habermasian species ethic of communication presupposes the intent of all parties in the debate to enter into a rational debate that leads to emancipatory knowledge through the adoption of a 'public' role that may differ from one's privately or culturally held beliefs.

The issue of understanding a new and complex science not only exposes the heart of the nanodebate, namely, the issue of education, but also suggests that the educative process of ethical dialogue provides the answer to the question. The proceduralist model is both method and *telos*.

The education process through which the topic of informed citizenry is debated would also potentially lead to some interesting insights into how global agents' skills might be better developed from a nano pTA process.

References

Aboulafia M (2001) The cosmopolitan self. George Herbert Mead and continental philosophy. University of Illinois Press, Urbana

Åm H (2011) Trust as glue in nanotechnology governance networks. NanoEthics 5(1):115–128, April

Anderson W (1997) The future of the self. Inventing the postmodern person. Putnam, New York

Andorno R (2009) Human dignity and global rights as a common ground for a global bioethics. J Med Philos 34:223–240

Apel KO (2008) Globalisation and the need for universal ethics. In: Cortina A, Garcia-Marquez D, Connill J (eds) Public reason and applied ethics. The ways of practical reason in a pluralist society. Ashgate, Aldershot, pp 135–154

Appiah KA (1996) Color conscious. Princeton University Press, Princeton

Appiah KA (2006) Cosmpolitanism. Ethics in a world of strangers. Allen Lane, London

Barber B (1996) Jihad vs McWorld: how globalism and tribalism are reshaping the world. Ballantine, New York

Bauman Z (2000) Liquid modernity. Polity Press, Cambridge

Beck U (2002) The cosmopolitan society and its enemies. Theory Cult Soc 19(1–2):12–44

Beck U, Grande E (2010) Varieties of second modernity: the cosmopolitan turn in social and political theory and research. Br J Sociol 61(2):410–443

Beck U, Sznaider N (2006) Unpacking cosmopolitanism for the social sciences: a research agenda. Br J Sociol 57(1):381–403 http://www2.mta.ac.il/~natan/unpacking%20cosmoplitanism%20 for%20the%20social%20sciences.pdf. Accessed 4 Jan 2015

Begley S (2010) East Brain, West Brain? What a difference culture makes. Newsweek, 18 Feb

Beitz CR (1994) Cosmopolitan liberalism and the state system. In: Brown C (ed) Political restructuring in Europe. Routledge, London, pp 123–136

Benhabib S (1993) Afterword. In: Benhabib S, Dallmyr RM (eds) The communicative ethics controversy. MIT Press, Cambridge, pp 330–359

Benhabib S (2006) Another cosmopolitanism. The Berkeley tanner lectures on human values. Oxford University Press, Oxford

Benn P (2001) Ethics. Taylor & Francis e-Library, London

Brandt R (2001) Ethical relativism. In: Moser PK, Carson TL (eds) Moral relativism: a reader. Oxford University Press, New York, pp 25–31

Calhoun C (ed) (1992) Habermas and the public sphere. MIT Press, Cambridge

Caney S (2005) Justice beyond borders: a global political theory. Oxford University Press, New York

Chomsky N (2002) A world without war. Delivered at the World Social Forum, 31 January 2002. http://www.chomsky.info/talks/200202--.htm. Accessed 4 Jan 2015

Cohen R, Vertovec S (2002) Conceiving cosmopolitanism. Theory, context, and practice. Oxford University Press, Oxford

Cook JW (1999) Morality and cultural differences. Oxford University Press, New York

Cortina A (2008) 'The public task of applied ethics: transnational civic ethics. In: Cortina A, Garcia-Marquez D, Connill J (eds) Public reason and applied ethics. The ways of practical reason in a pluralist society. Ashgate, Aldershot, pp 9–32

Cortsese AJ (1990) The restructuring of moral theory. SUNY Press, New York

Davies G (2007) Habermas in China: theory as catalyst. China J 57(January):61–85

De Castro LD (1999) Is there an Asian bioethics? Bioethics 13:227–235

DeCew JW (1990) Moral conflicts and ethical relativism. Ethics 101(1):27–41

Delanty G (2000) Citizenship in a global age. Society, culture, politics. Open University Press, Buckingham

Derrida J (2001) On cosmopolitanism and forgiveness. Routledge, London

Dhanda M (2008) The negotiation of personal identity. Verlag Dr Muller, Saarbrucken

Dien DS (1983) A Chinese perspective on Kohlberg's theory of moral development. Dev Rev 2:331–341

Dower N (2005) Situating global citizenship. In: Germain RD, Kenny M (eds) The idea of global civil society. Politics and ethics in a globalizing era. Routledge, Abingdon, pp 100–118

Dower N (2007) World ethics. The New Agenda. Edinburgh University Press, Edinburgh

Etzioni A (1996) A moderate communitarian proposal. Polit Theory 24(2):155–171

Etzioni A (2002) Are particularistic obligations justified? A communitarian examination. Rev Polit 64(4):573–598

Eze MO (2011) I am because you are. The UNESCO Courier, October–December 2011, pp 10–14

Fairfield P (1999) Hermeneutical ethical theory. In: Madison GB, Fairbairn M (eds) The ethics of postmodernity: current trends in continental thought. Northwestern University Press, Evanston, pp 138–162

Featherstone M (2002) Cosmopolis. An introduction. Theory Cult Soc 19(1–2):1–16

Ferrara A (1990) Universalism: procedural, contextualist and prudential. In: Rasmussen D (ed) Universalism vs. communitarianism. Contemporary debates in ethics. MIT Press, Cambridge, pp 11–38

Finlayson JG (2005) Habermas. A very short introduction. Oxford University Press, Oxford

Fung ESK (2006) The idea of freedom in modern China revisited: plural conceptions and dual responsibilities. Mod China 32:453–482

Gadamer HG (2004) Truth and method. Continuum International Publishing Group, London

Gan S (2012) The destiny of modern virtue ethics. Front Philos China 5(3):423–448

Gergen K (1990) The saturated self: dilemmas of identity in contemporary life. Basic, New York

Gill C (2005) In what sense are ancient ethical norms universal? In: Gill C (ed) Issues in ancient and modern ethics. Clarendon, Oxford, pp 15–40

Gilligan C (1982) See her in a different voice: psychological theory and women's development. Harvard University Press, Cambridge, MA

Gurtov M (1994) Global politics in the human interest. Lynne Rienner Publishers, Boulder, pp 6–11

Gutmann A (1993) The challenge of multiculturalism in political ethics. Philos Public Aff 22(3):171–206

Gutmman A, Thompson D (1996) Democracy and disagreement. Bellknap/Harvard University Press, Cambridge

Habermas J (1979) Communication and society. Heinemann, London

Habermas J (1991) Moral consciousness and communicative action. MIT Press, Cambridge

Habermas J (1992a) Postmetaphysical thinking: philosophical essays (trans: Hohengarten WM). MIT Press, Cambridge

Habermas J (1992b) Discourse ethics, law and Sittlichkeit. In: Dews P (ed) Autonomy and solidarity: interviews with Jürgen Habermas. Verso, London, pp 245–271

Habermas J (1993) Justification and application: remarks on discourse ethics (trans: Cronin C). MIT Press, Cambridge

Habermas J (1998a) Between facts and norms. MIT Press, Cambridge

Habermas J (1998b) The inclusion of the other. Studies in political theory. MIT Press, Cambridge

Habermas J (2002) Wann müssen wir tolerant sein? Über die Konkurrenz von Weltbildern, Werten und Theorien. Lecture at the Leibniz Conference, Berlin-Brandenburg Academy of the Sciences, 29 June 2002

Habermas J (2003) The future of human nature (trans: Beister H, Rehg W). Polity Press, Cambridge

Habermas J (2004) The theory of communicative action, vol 2, Lifeworld and system: a critique of functionalist reason. Polity Press, Cambridge

Hendley S (2000) From communicative action to the face of the other. Levinas and Habermas on language. Obligation and community. Lexington Books, Lanham

Hursthouse R (1999) On virtue ethics. Oxford University Press, Oxford

Hutchings K (2010) Global ethics. An introduction. Polity, Cambridge

Jameson F (1988) Cognitive mapping. In: Nelson C, Grossberg L (eds) Marxism and the interpretation of culture. University of Illinois Press, Chicago, pp 347–360

Jonas H (1984) The imperative of responsibility. University of Chicago Press, Chicago

Jordaan E (2009) Dialogic cosmopolitanism and global justice. Int Stud Rev 11:736–748

Jotterand F (2006) The politicization of science and technology: its implications for nanotechnology. J Law Med Ethics (Winter): 658–666

Kessler CS (2000) Globalization: another false universalism? Third World Quart 21(6):931–942

Keung H (1988) The Chinese perspectives on moral judgment development. Int J Psychol 23:201–227

Kim YY (2008) Intercultural personhood: globalization and a way of being. Int J Intercult Relat 32(4):359–368

Kjølberg KL, Strand R (2011) Conversations about responsible nanoresearch. NanoEthics 5(1):99–113, April

Klein N (2000) No logo. Flamingo, London

Korsgaard C (1996) Self-constitution, agency, and identity. Oxford University Press, Oxford

Korsgaard C, Cohen GA, Geuss R, Nagel T, Williams B (2009) The sources of normativity. Cambridge University Press, Cambridge

Kymlicka W (2007) The globlization of ethics. In: Sullivan WM, Kymlicka W (eds) The globalization of ethics. Cambridge University Press, Cambridge, pp 1–16

Lee S (2001) Transversal-universals in discourse ethics: towards a reconcilable ethics between universalism and communitarianism. Human Stud 24(1–2):45–56

Lehrer K (2001) Individualism, communitarianism and consensus. J Ethics 5(2):105–120

Liang T (2009) Mencius and the tradition of articulating human nature in terms of growth. Front Philos China 4(2):180–197

Lukes S (2008) Moral relativism. Profile, London

MacIntyre A (1984) After virtue. A study in moral theory. University of Notre Dame Press, Notre Dame

Mackie J (1990) Ethics: inventing rights and wrongs. Penguin, London

Madison GB (1997) The ethics and politics of the flesh. In: Madison GB, Fairburn M (eds) The ethics of postmodernity. Current trends in continental thought, vol Il. Northwestern University Press, Evanston, pp 174–190

Madison GB (2002) China in a globalizing world: reconciling the universal with the particular. Dialogue Univers 11–12:51–79

Maharaj S (2010) Small change of the universal: beyond modernity? Br J Sociol 61(3):565–578

Mann J, Gruskin S, Grodin M, Annas G (1999) Health and human rights. Routledge, New York

McCarthy T (1978) The critical theory of Jürgen Habermas. MIT Press, Cambridge

McCarthy T (1991) Ideals and illusions. MIT Press, Cambridge

McCarthy J (2003) Principlism or narrative ethics: must we choose between them? Med Humanit 29(2):65–71

Mead GH (1982) The individual and the social self. University of Chicago Press, Chicago

Mead GH, Morris CW (1934) Mind, self & society: from the standpoint of a social behaviourist. University of Chicago Press, Chicago

Merleau-Ponty M (1962) Phenomenology of perception (trans: Smith C). Routledge, London

Miller D (1998) The limits of cosmopolitan justice. In: Mapel DR, Nardin T (eds) International society. Diverse ethical perspectives. Princeton University Press, Princeton

Miller D (2002) Equality and justice. In: Chadwick R, Schroeder D (eds) Applied ethics, vol 6, Critical concepts in philosophy. Routledge, London, pp 231–254

Morin ME (2009) Cohabiting in the globalised world: Peter Sloterdijk's global foams and Bruno Latour's cosmopolitics. Soc Space 27:58–72

Nagel T (1979) The possibility of altruism. Princeton University Press, Princeton

Neem JN (2009) The universe in a grain of sand? Cosmopolitanism and nationalism reconsidered: a review of Seyla Benhabib's another cosmopolitanism, Kwame Anthony Appiah's cosmopolitanism, and Craig Calhoun's nations matter. Hedgehog Rev (Fall):51–58

Ng SH, Han S, Mao L, Lai JCL (2010) Dynamic bicultural brains: fMRI study of their flexible neural presentation of self and significant others in response to culture prime. Asian J Soc Psychol 12:83–91

Nussbaum M (1996) Patriotism and cosmopolitanism. In: Cohen J (ed) For love of country: debating the limits of patriotism. Beacon, Boston, pp 2–17

Nussbaum M (2010) Veiled threats. Opinionator, The New York Times, July 11. http://opinionator. blogs.nytimes.com/2010/07/11/veiled-threats/?_r=0. Accessed 4 Jan 2015

O'Neill O (2001) Bounded and cosmopolitan justice. In: Booth K, Dunne M, Cox T (eds) How might we live? Global ethics in a new century. Cambridge University Press, Cambridge, pp 45–60

Okin SM, Cohen J, Howard M, Nussbaum M (1999) Is multiculturalism bad for women? Princeton University Press, Princeton

Pellizzioni L (2004) Responsibility and environmental governance. Environ Polit 13:541–565

Philips M (1994) Between universalism and skepticism. Ethics as social artifact. Oxford University Press, New York

Philpott D (2007) Global ethics and the international law tradition. In: Sullivan WM, Kymlicka W (eds) The globalization of ethics. Cambridge University Press, Cambridge, pp 17–37

Pinker S (1994) The language instinct. W. Morrow & Co, Berkeley

Pojman LP (2005) Kant's perpetual peace and cosmopolitanism. J Soc Philos 36(1). http://onlinelibrary.wiley.com/doi/10.1111/j.1467-9833.2005.00258.x/pdf. Accessed 4 Jan 2015

Ricoeur P (1992) Oneself as another (trans: Blamey K). University of Chicago Press, Chicago

Ricoeur P (2005) The course of recognition (trans: Pellauer D). Harvard University Press, Cambridge

Robbins B (1998) Introduction part one: actually existing cosmopolitanism. In: Cheng P, Robbins B (eds) Cosmopolitics: thinking and feeling beyond the nation. University of Minnesota Press, Minneapolis, pp 1–19

Robeyns I (2002) In defence of Amartya Sen. Post-autistic economics review 17(5). http://www. btinternet.com/~pae_news/review/issue17. Accessed 4 Jan 2015

Rolbin C, della Chiesa B (2010) We share the same biology. Cultivating cross-cultural empathy and global ethics through multilingualism. Mind Brain Educ 4(4):196–207

Rorty R (1980) Philosophy and the mirror of nature. Blackwell, Oxford

Rorty R (1989) Contingency, irony, and solidarity. Cambridge University Press, Cambridge

Rousseau JJ (1986) The first and second discourses (trans: Gurevich V). Harper & Row, New York

Rumelili B (2008) Interstate community-building and the identity/difference predicament. In: Price RM (ed) Moral limit and possibility in world politics. Cambridge University Press, Cambridge, pp 253–380

Rumford C (ed) (2007) Cosmopolitanism and Europe. Liverpool University Press, Liverpool

Sakamoto H (1999) Towards a new "global bioethics". Bioethics 13:191–197

Sandel M (1998) Liberalism and the limits of justice. Cambridge University Press, New York

Scanlon TM (1982) Contractualism and utilitarianism. In: Sen A, Williams B (eds) Utilitarianism and beyond. Cambridge University Press, Cambridge, pp 103–128

Scherer AG, Patzer M (2011) Beyond universalism and relativism: Habermas' contribution to discourse ethics and its implications for intercultural ethics and organization theory. Res Soc Org 32:155–180

Schneider HJ (1994) Ethisches argumentieren. In: Martens E, Hastedt H (eds) Ethik. Ein Grundkurs. Rowohlt, Hamburg, pp 13–47

Scott CE (1999) The sense of transcendence and the question of ethics. In: Madison GB, Fairbairn M (eds) The ethics of postmodernity. Current trends in continental thought. Northwestern University Press, Evanston, pp 214–230

Sen A (1987) The standard of living. Cambridge University Press, Cambridge

Sen A (1999) Development as freedom. Oxford University Press, Oxford

Sen A, Williams B (eds) (1982) Utilitarianism and beyond. Cambridge University Press, Cambridge

Simpson E, Williams M (1999) Reconstructing Rorty's ethics: styles, languages, and vocabularies of moral deliberation. In: Madison GB, Fairbairn M (eds) The ethics of postmodernity. Current trends in continental thought. Northwestern University Press, Evanston, pp 120–137

Singer P (1972) Famine, affluence, and morality. Philos Public Affairs 1(1) (Spring):229–243. http://www.utilitarian.net/singer/by/1972----.htm. Accessed 4 Jan 2015

Solomon RC (2006) On ethics and living well. Thomson/Wadsworth Philosophical Topics, Belmont

Stanford Encyclopedia of Philosophy (2001) Communitarianism. http://plato.stanford.edu/entries/communitarianism. Accessed 4 Jan 2015

Stanford Encyclopedia of Philosophy (2002) Cosmopolitanism. http://plato.stanford.edu/entries/cosmopolitanism/#2. Accessed 4 Jan 2015

Stanford Encyclopedia of Philosophy (2007) Jürgen Habermas. http://plato.stanford.edu/entries/Habermas. Accessed 4 Jan 2015

Stiglitz J (2003) Globalization and its discontents. W.W. Norton, New York

Subra S (2011) Moving towards global science. Science 333(6044):802

Szerszynski B (1999) Risk and trust: the performative dimensions. Environ Value 8:239–252

Taylor C (1992) Multiculturalism, and "the politics of recognition". Princeton University Press, Princeton

Taylor C (1995) Philosophical arguments. Harvard University Press, Cambridge

Tehranian M (1999) Global communication and world politics. Lynne Rienner Publishers, Boulder

Therborn G (2000) At the birth of second century sociology: times of reflexivity, spaces, and nodes of knowledge. Br J Sociol 51(1):37–57

Todorov T (1993) On human diversity. Nationalism, racism, and exoticism in French thought (trans: Porter C). Harvard University Press, Cambridge

Tran VD, Shen V (1991) Chinese foundations for moral education and character development, vol III 2. CRVP Press, Columbia

Turner B (2002) Cosmopolitan virtue, globalization and patriotism. Theory Cult Soc 19(1):45–73

Tutu D (2012) We are all of one family. The Guardian, 22 October

Venn C (2002) Altered states: post-enlightenment cosmopolitanism and transmodern socialities. Theory Cult Soc 19(1–2):65–80

Von Goethe JW (1972) West-ostlicher Divan (1814-19). Suhrkamp, Frankfurt

von Schomberg R (2007) From the ethics of technology towards an ethics of knowledge policy and knowledge assessment. EC Directorate-General, Brussels. http://ec.europa.eu/research/science.../ethicsofknowledgepolicy_en.pdf. Accessed 4 Jan 2015

Walby S (2002) From community to coalition: the politics of recognition as the handmaiden of the politics of equality in an era of globalization. In: Lash S, Featherstone M (eds) Recognition and difference. Sage, London, pp 113–136

Warnke G (1993) Justice and interpretation. MIT Press, Cambridge

Watt AJ (1975) Transcendental arguments and moral principles. Philos Quart 25:40–57

Webb SA (2006) Social work in a risk society. Palgrave Macmillan, Basingstoke

Weisband E (2007) Conclusion: a postmodern public ethics. In: Ebrahim A, Weisband E (eds) Participation, pluralism, and public ethics. Cambridge University Press, Cambridge, pp 308–309

Werbner P (1999) Global pathways. Working class cosmopolitans and the creation of transnational ethnic worlds. Soc Anthropol 7(1):17–35

Westwood S (2002) Complex choreography: politics and regimes of recognition. In: Lash S, Featherstone M (eds) Recognition and difference. Sage, London, pp 247–264

White SK (1988) Ethics, politics and history: an interview with Jorgen Habermas, conducted by J-F. Ferry. Philos Soc Crit 14:433–439

Widdows H (2007) Is global ethics moral neo-colonialism? An investigation of the issue in the context of bioethics. Bioethics 6(21):305–315

Wisnewski JJ (2008) The politics of agency. Towards a pragmatic approach to philosophical anthropology. Ashgate, Aldershot

Wittgenstein L (1965) A lecture on ethics. Philos Rev 74:3–12

Wohlrapp H (1998) Constructivist anthropology and cultural pluralism: methodological reflections on cultural integration. In: Kumar BN, Steinmann H (eds) Ethics in international management. Walter de Gruyter, Berlin, pp 47–63

Wolfe S (1992) Two levels of pluralism. Ethics 102(4):785–798

Yang H (2012) Universal values and Chinese traditional ethics. J Int Bus Ethics 3(1):81–90

Yu SJW (2003) 'Virtue: confucius and Aristotle. In: Xiang J (ed) The examined life: the Chinese perspective. Global Publications, Binghampton

Zeiler K (2009) Self and other in global bioethics: critical hermeneutics and the example of different death concepts. Med Health Care Philos 12:137–145

Zolo D (1997) Cosmopolis: prospects for world government. Polity Press, Cambridge

Chapter 8
Conclusion: Discourse Ethics and the Dialectics of East-West Intersubjectivity

> *I call it a draw. No decisive choice should be made between universalism and particularism. (Callahan 2000)*
>
> *'Only connect!' (Forster 1910)*

This work started with an interest in three seemingly divergent topics: global ethics (a topic of interest to me given my work in a multicultural setting), nanotechnology (about which I knew a bit from avid reading of science fiction novels), and virtue ethics. While at first it seemed that these three things had little in common, I came to believe quite the opposite.

Global ethics – or rather the lack thereof – seems to be a problem to be dealt with at a time when the globe is rapidly forging ahead with a new technology, one that may be riskier than we currently know. There seems little point in developing a country-specific approach to a technology that transcends boundaries, and thus I began with the choice of two divergent regions, China and the EU, which I assumed (following a commonly held view) to be very different in their cultural contexts as well as their approaches to S&T regulation. As there had been very little written on Chinese nanoethics, it seemed an interesting challenge and a useful contribution to the field.

Nanotechnology has been presented as a key site for experimenting with novel forms of so-called upstream engagement or efforts to engage members of the public in dialogue about emerging technologies (ENSAA 2011). It provides a context for the emergence of a new risk governance paradigm, in terms of which political culture and risk perception in local societies are becoming crucial factors in risk assessment and governance (Chou 2007). In asking whether Eastern and Western approaches to nanotechnology governance can be aligned, one can observe that Europe is increasingly cooperating and competing with both China and India, which are also keen to develop their S&T sectors. Such new interdependencies between global actors require new global approaches to S&T policy or at least the recognition of differing local approaches to global science.

The first task was to see if China and the EU were so very different in their views on nanotechnology – the answer being yes and no. Chapters 1 and 2 concluded that the bioethics context (unsurprisingly) differs in the East and the West. A brief look at various guidelines, for example, in the context of cloning, suggests that as person-

© Springer International Publishing Switzerland 2015

S. Dalton-Brown, *Nanotechnology and Ethical Governance in the European Union and China*, DOI 10.1007/978-3-319-18233-9_8

hood can only be acquired through social practice, according to Confucian teachings, human value evolves out from an individual's social relations. Western individualism would take a different approach, one based more on the right to individual autonomy. This outlines the basic conflict as it is often perceived in broad terms between West and East. However, Chap. 2 argued the point that 'dignity', seen as a primary value in Western discourse on the individual and his or her rights, is a more universal concept than is usually thought. The so-called clash of civilisations is often seen as a clash of individualism, meaning that the Western stress on human rights is irreconcilable with Eastern communitarianism, which places the good of society over individual rights. A more useful way to approach this issue is to look at dignity as more fundamental to Western individualism than political notions of human rights. Dignity should be seen as a form, not even of autonomy, as that too has connotations of human rights, but rather of virtuous agency. It is a process, not a values-based system.

Related conclusions were:

(a) In terms of nanosafety as expressed through policy and regulation, China and the EU have similar approaches towards and concerns about nanotoxicity – the official debate on benefits and risks is not markedly different in the two regions.
(b) That there is a similar economic drive behind both regions' approaches to nanodevelopment, the difference being the degree of public concern admitted.
(c) That – most significantly – participation in decision-making is fundamentally different in the two regions.

Reading about nanotechnology issues reveals that risk is paramount, but also that risk is not dealt with in the same way by each region, as indicated by their nanopolicies (the focus of Chap. 3). The precautionary principle in the EU and economic drivers in China make for a difference in emphasis that translates, obviously enough, into the way the public has been involved in S&T analysis. Here the interesting fact is that issues of public perception have been emerging in China; might the region approach participatory technology assessment differently from the West? Here I discovered that there has not been much pTA in China worth speaking of. Thus, in China, the focus is on the responsibility of the scientist; in the EU, it is about government accountability to the public. This may change in China, since individual responsibility alone cannot guide S&T development and as public participation is increasingly seen as integral to governmental decision-making more globally.

Given increasing public concern, post-GM food, about far-reaching technologies, pTA would seem to be the logical method for further discussion of potentially global approaches to nanotechnology. Whereas the official debate on benefits and risks is not markedly different in both areas, the public debate in China lags behind the EU. This is partly due to the fact that the public in China currently appears more concerned about GM food as well as the lack of channels for public participation in China. Yet public participation is globally seen as increasingly integral to governmental decision-making, particularly given the economic effects of product boycotting, and so China might soon (and perhaps already is in terms of GM food) face this inevitable issue. The two regions are converging.

This occasioned two questions – is pTA so important, and if it is, how might it be done most effectively worldwide? Reading about pTA was to take me into a variety of eclectic environments, as the business world has become engaged with the idea of community or stakeholder involvement, while sociologists, political philosophers, and communication ethics writers have all come to realise the importance of community engagement for successful product implementation (to put it rather cynically). This also led me to Habermas, whose work seems to provide a much richer theoretical background to pTA than commentators (Leo Hennen's recent work being one exception) had considered and which seemed to me to expose a deficit in discussions of nanofora.

Habermas, although he has his detractors, has a few sensible ideas for anyone looking at global ethics and public engagement (or as he would put it, deliberative democracy, which he sees as a legitimising force for initiatives such as new technology development). In particular, Habermas is very keen on the idea of polyphonic discourse, one in which consensus is achieved through the recognition of differing viewpoints. This seemed to me to be the key question of global ethics – how might citizens come together and surmount cultural differences so as to agree on what the world might like to do with potentially revolutionary technology?

This is an idealistic question, though it is one which I felt could be answered pragmatically. In fact, a pragmatic approach seemed to be the solution, particularly after some reading on the ongoing debate between proponents of universal values, and those of relative or culturally determined values, which revealed how discussions like this would always be tricky. The *discussion,* however, was the important part of the statement, as from dialogue one might develop a methodology that would work towards consensus through procedure rather than content. In short, it should be a rational discourse, as Habermas suggests, aimed towards a global ethics methodology.

There has been some work done on this topic, ranging from discussions about the role of the public in offering new and creative approaches to the field to whether the nanoethics field requires a radical new paradigm, to work on narrative and the role of the humanities in offering a multidisciplinary and thus creative methodology, to the idea of a clash between references to the technological past and suppositions about its future. Such work, while it clarifies the issues that are being discussed, and on what basis this takes place – i.e. whether they derive from arguments based on a consequentialist position or arguments on human dignity and autonomy – seemed to circle a particular issue concerning what the best practice might look like if we were to achieve consensus.

This is the moment when I realised that Aristotle did have a point to make. If global ethics in the nanotechnology field requires public input (upstream engagement, as it is called), then it seemed odd to ignore the nature of the global agent, whose task would be to participate in public debate and perhaps to develop creative approaches to nanoethics.

Nanoethics agency is not often analysed by critics, though there has been an emerging realisation that the more familiar consequentialist and deontological positions might be usefully engaged with that of virtue ethics. Comparisons have been made between Habermas and Aristotle, for example, in McIntyre's work on their political views. However, there has been little on the idea of an Aristotelian-Habermasian global agent who would use the ideas of Chinese commentators. Interweaving such strands into the new work on Chinese pTA, Habermas's views of intercultural dialogue, and the skills required by a practical discourse agent seemed to me to offer a potential basis for a new approach to the nanoethics debate.

Global ethics is to be achieved by global agents, who are not only policymakers and scientists but laypersons as well. Looking at pTA as a useful focus for points of similarity or difference between the Eastern and Western approaches to ethics, it seemed logical to turn to the work of Jürgen Habermas, whose comments on discourse ethics bear obvious similarities to the pTA process, particularly in his emphasis on discourse as necessarily inclusive and multi-voiced (see Chap. 4). Habermas has also been relevant to the emphasis placed in this study on procedure and agency. The latter, with its connection to Amartya Sen's focus on what people are 'capable' of, provides a practical basis for discourse and international dialogue.

Habermas asks a basic question of global ethics, namely, how different views (particularly of social order) can be universally recognised and reconciled, perhaps within an 'ideal community' of communication that may be global? Discourse ethics focuses on what is 'equally good for all', action-related conflicts being resolved with reference to a generalisable interest (Habermas 1993). This is similar to the practical statement by Ladikas and Schroeder (2005) that global ethics 'is not a field of academic study, it is an activity; the attempt to agree on fundamental conditions for human flourishing and to actively secure them for all.' It is reliant on such practical fora as platforms for intercultural dialogue and trust-building, as well as international ethics committees and ethics reviews for ongoing global negotiations.

As Chap. 6 argues, while agents of discourse are required to discuss the consequences of new technology and to examine the morality of the technological activity, the main aim for such agents is that of achieving rational dialogue. To do this, they need to possess those virtues that impel them to enter fully into such intersubjective dialogue. The significant activity or agency is the key to this model of discourse ethics/pTA.

Habermas's orientation towards reasoning, rather than rationality, is examined with a view to arguing that as Habermasian thought (e.g. particularly as developed by Benhabib) incorporates an element of care ethics, there is a point of crossover with Confucian *ren* (benevolence or kindness) ethics.

Habermas's view of emotion relates to procedural issues, such as coming to dialogue with a sincere and committed mindset, one that allows for empathy or 'considerateness.' *Ren* ethics seems fairly similar to the Habermasian species ethics on the following points:

(a) It reveals itself in a relational form.
(b) It requires sincerity or 'inner form'.
(c) And it is linked to ethical maturation – a concept Habermas has discussed in his view of human development and how the human being progresses through various stages of knowledge.

However, even if there is a similarity to how agents approach discourse, this is not to say that their 'knowledge form' (mode of argumentation) is alike. Against so-called Western rational individualism, one might predicate the reputed Asian ideal of harmonious collectivism, raising the issue of whether the Habermasian model can adapt to a less rationally teleological and less autonomous decision-making environment.

Having reached this point, it struck me that the work of this study was only three-quarters complete, and despite my interest in the fields of nanotechnology, science fiction novels, bioethics, the GM debate, the differences between Eastern and Western views of dignity, Aristotle, Habermas, deliberate democracy, pTA, communication ethics, and so forth, I needed to look further to define what exactly the global agent should *do*. Embracing further eclecticism in my reading, I considered the issue of the global agent's process of debate – what stages might he or she go through in the attempt to achieve consensus and overcome any culturally determined viewpoints that might prevent consensus? Of course, this is as broad a question as asking, 'why can't we get along'?

An answer to this problem is potentially supplied by my development of a five-stage model of identity formation for the global agent, one that emerges from considering the works of various critics such as Taylor. The model utilises the following five steps:

- Decentralisation – A process intended to ensure a neutral debate
- Recognition – A process intended to ensure that all interests and viewpoints are recognised
- Reciprocity – A process intended to recognise the value of viewpoints with which we do not agree
- Negotiation – Subjecting those viewpoints to (Habermasian) rational debate
- Identity formation – Achieving consensus by adopting a dual identity through intersubjectivity and dialectic

This model offers a synthesis of ideas on cultural identity while contextualising them within a pTA/Habermasian process that hopefully could be of use in dialogue processes.

The Asian (Confucian virtuous agent) is similar to the Aristotelian and the Habermasian in that:

(a) Both 'saw moral virtues as . . . derived not from a universal moral calculus but from a careful process of personal discovery' (Gier 2004).
(b) Both have an intersubjective view of ethics based on a dialectic relationship between self and others.

Intersubjectivity is a key concept for any theory of global ethics. The basis for supporting such an applied model derives from the concept of moral identity, the formation of which requires one to go beyond one's 'context', for 'universalistic action orientations ... make it possible to gain some distance from the social roles that shape one's background and character' (Habermas 2004); Habermas (1992) suggests ego identity as a dual concept, reflecting an interdependence of society and one's inner self; one can both understand the cultural biases inherent in any act of communication while acting autonomously, free from such bias. In terms of dialecticism, it implies one's socially circumscribed self is also in a productive tension with a public or broader self. For the Western discourse agent, this means more community focus, for the Eastern, more individualism. Thus the pTA/Habermasian model of intersubjective ethics requires a dialectic between self as individual and self as other or self as collective.

It turns the usual negative of East-Western ethics (the individual and the collective are two very foreign notions) into a positive (the dialectic between individual and collective is what makes ethics work).

There is a further meaning of the word 'dialectic.' A distinction, between dialogue and dialectic, suggests that in order to achieve a state of agency in which one can act as a global citizen, one needs to achieve an intersubjective state. Such intersubjectivity can be achieved through dialectic. Dialectical agency implies the validation of one's words only through the speech of the other. This suggestion, based on a Bakhtinian approach, is a new suggestion in work on how nanoethics might develop on the basis of procedural agency (pTA).

There is a key methodological issue – whether the East-West real difference may be stated in terms of the dialectic process implied by the Habermasian model, in which self and other exist in a dialogic state. The Confucian one implies that such a dialogic or dialectic state must operate towards the goal of social harmony, the Habermasian towards the goal of consensus. There appears to be a *difference in terms of how much dialecticism is allowed,* in that Western individualism encourages greater opposition to the whole.

The argument runs thus: every statement contains in fact the echo of its own opposition, since speech is inherently oppositional. Thus in discourse, one both says what is true to oneself, but also predicates the echo of an opposing view, the voice of the other. This implies that the discourse agent, to achieve emancipatory knowledge, must exist within a dialectic, or the tension between relative and universal values, between the goal of the common good, and that of individual and contextual desires. Whether she or he can maintain that tension or not might be the real issue of global ethics.

References

Callahan D (2000) Universalism and particularism. Fighting to a draw. Hastings Centre Rep 30(1):37–44

Chou KT (2007) Biomedtech Island Project and Risk Governance. Paradigm conflicts with a hidden and delayed high-tech society. Soziale Welt 58:123–143

ENSAA (2011) Nanotechnology and society: moving upstream! http://www.ensaa.eu/index.php/innovation/106-nanotechnology-moving-upstream.html. Accessed 14 Jan 2015

Forster EM (1910) Howard's end. Pover Thrift, London

Gier NF (2004) Whitehead, confucius, and the aesthetics of virtue. Asian Philos 14(2):171–190

Habermas J (1992) Discourse ethics, law and Sittlichkeit. In: Dews P (ed) Autonomy and solidarity: interviews with Jürgen Habermas. Verso, London, pp 245–271

Habermas J (1993) Justification and application: remarks on discourse ethics (trans: Cronin C). MIT Press, Cambridge

Habermas J (2004) The theory of communicative action, vol 2, Lifeworld and system: critique of functionalist reason (trans: McCarthy T). Polity, Cambridge

Ladikas M, Schroeder D (2005) Too early for global ethics? Camb Q Health Care Ethics 14(4):404–410

Appendix: African Nanoethics

> Peoples of African descent…are linked by shared values that are fundamental features of African identify and culture. These, for example, include hospitality, friendliness, the consensus and common framework-seeking principle, ubuntu, and the emphasis on community rather than on the individual. These features typically underpin the variations of African culture and identity everywhere. The existence of African identity is not in doubt. (Nyasani 1997)

Nanotechnology is certainly important to Africa. In 2005, the UN Millennium Project's Task Force on Science, Technology and Innovation had already identified the technology as an important tool for addressing poverty in Africa. The Millennium Goals are accompanied by initiatives such as the World Summit on Sustainable Development (WSSD), the Blair Commission Report, and the New Partnership for Africa's Development (NEPAD), all of which foreground science and technological innovation as major driving forces for African development. The nine member countries of the Common Market for Eastern and Southern Africa (COMESA), at a summit in 2010 on harnessing science and technology for development, urged the promotion and utilisation of nanotechnology and science, particularly given its application in various key areas such as medical treatment (University World News 2010).[1] The impact of nano on developing countries could in fact be radical, in that it could lead to (Schummer 2007; Invernizzi 2008):

1. Improved crop yields
2. Clean water
3. Access to more energy (improved energy storage)
4. Better packaging to improve the shelf life of food
5. Better construction materials

---

[1] Focus nanotechnology Africa Inc. (FONAI) was formed in 2006 as a South African-Nigerian joint not-for-profit educational and scientific organisation especially in the USA, Africa, and the Caribbean 'to combat brain drain and all forms of poverty including science and technological poverty'. It repeatedly pleads for donations and there is little information on its achievements.

© Springer International Publishing Switzerland 2015
S. Dalton-Brown, *Nanotechnology and Ethical Governance in the European Union and China*, DOI 10.1007/978-3-319-18233-9

6. More widely available health screening and better treatments (Acharya et al. 2004; Salamanca-Buentello et al. 2005)

So much for the positive, but in negative terms there are concerns about nanotechnology, with some African nations arguing for the right to 'say no to nano', i.e. to accept or reject the import and use of manufactured nanomaterials to minimize risk, and calling for more work to be done on nanotechnology's ethical and social risks to developing countries and countries with economies in transition' (Wetter 2010).

The issues, apart from the global concern with risk in terms of a relatively new technology, are principally those of the nanodivide and also of what might be termed risk colonisation. The United Nations Development Programme (UNDP) in its Human Development report in 2001 endorsed biotech as a means of improving food supply in developing countries, with the caveat that it should be introduced into such countries with a view to sustainability. The alternative is to 'utilise' Africa as a testing ground; one might note the USA's attempt to impose GM crops on southern African countries. As unmilled maize was included in the offering – meaning cross-fertilisation with local crops – the approach would have been disallowed in the EU, and the OECD took the view that it was unethical.[2]

There are other issues, such as economic upheaval, as well as the problem that manufacturing in countries with weaker controls (and subsequently exported worldwide may increase risk; and benefits may be unevenly distributed.

If nano is a 'GP' or general-purpose technology that impacts on the productivity of many technologies, there may be global job losses and an increasing move towards a workforce with nanoskills, in terms of which the developing world with its unskilled labour market may be disadvantaged (Valdivia 2011). New nanoproducts may also lead to major changes in trade balances between countries. Wetter (2010), writing on nanotechnology as not such a good idea for Africa, notes potential 'major disruptions to traditional commodity markets', for 'if a new nano-engineered material outperforms a conventional material and can be produced at a comparable cost, it is likely to replace the conventional commodity' and so critical export earners in developing countries, namely, raw materials, may be replaced by nanomaterials. And worker displacement brought on by commodity obsolescence 'will hurt the poorest and most vulnerable, particularly those workers who don't have the economic flexibility to respond to sudden demands for new skills or different raw materials' (Wetter 2010).

However, the nanodivide is Africa's chief *ethical* concern. The combination of visionary research with possible (even if to some, unfeasibly so) enormous societal, medical and environmental impact, as well as lucrative short- and long-term benefits encouraging a rush to market, does indeed sound like an uneasy mix for African states concerned (perhaps idealistically) about whether profits or world problems might be more important (Hermerén 2007). As Hunt states (2008), 'can we at last

---

[2] Unsurprisingly, the USA has refused to sign up to the Cartagena Protocol (2003) that governs trade in Living Modified Organisms.

know ourselves well enough to make an international cooperative effort to put nano-technological developments at the service of human and ecological welfare, or will it be primarily nanotechnology for more over-consumption?' In an ideal world, all nano R&D would be subject to the Millennium Goals, particularly those of eradicating poverty, hunger, and health issues; all governments and corporations would support a global fund (somewhat like Thomas Pogge's Health Impact Fund), their aims being neither those of political piggybacking on popular national initiatives nor of short-term profiteering. In the harsh realities of today's global commerce, nanoengineered tennis balls will still be more profitable than developing improved and ultra-cheap water filtration units for Africa. There is also the issue of corporations wishing to recoup research outlay by maintaining monopolies on products: 'Without targeted action, it is likely that many of the benefits nanotechnology can provide to the developing world will be delayed by at least a generation or more – the 20-year term of a patent' (Heller and Peterson 2007). Wetter (2010) argues that researchers in the South are likely to find that participation in the proprietary nanotech revolution is 'highly restricted by patent tollbooths, obliging them to pay royalties and licensing fees to gain access…nanotech will profoundly affect Africa's economy'.

In pharmaceutics this is particularly clear, as ongoing debates over generic drugs versus their more expensive patented versions and of the 'values gap' between populations that can pay for drugs and those who cannot have shown.[3] The developing world may find itself unable to fund sufficient research to develop its own products, and reduced to the status of a global market for developed world corporations (Macklin 2004; Schummer 2007; Jamison 2009). Due to research funding and the rush to patent, one may find a concentration of the power of nanotechnology in the hands of few transnational corporations headquartered in developed countries, something that 'goes hand in hand with inequity' in the view of Foladori et al. (2009). Such corporations, they claim, are intent on:

> …the control of risks as long as they do not imply less profit; the adoption of voluntary codes of conduct rather than mandatory regulations implemented by governments; the guarantee of health and environmental security only after obtaining technological developments to deal with these issues; and all these only after allocating nanotechnologies in the market.

Nanotechnology arguably (as with any high-tech industry) enters a situation of existing and widening divides. The 'nanodivide' between developed and developing countries (often called the North-South divide) can mean the gap between the 'information rich' and the 'information poor' but also can refer to inequity based on the (more profitable) areas towards which nanotechnology research is targeted, as compared to the areas in which it would address basic human needs. Maclurcan (2010) notes that there is 'little consideration' of 'what Southern populations might lose through trade liberalisation', with Southern countries having little recourse for protecting their markets or entering the race for 'land grab' for nanotechnology

---

[3] 'The question of the social good and to whom it applies filters through every phase of pharmaceutical production'. See Petryna et al. (2006).

patenting, one that 'far surpasses what was seen in the equivalent historical period for biotechnology patenting'.

The South's role as a market for nanoproducts is also potentially enormous, as is its role in offering a lower-cost manufacturing environment, so there is little incentive to support North-South start-ups. Given the embryonic status and long-term potential payback of many of nano's stages of development, the assumption is that much of the research would be too costly in terms of human capital and infrastructure to do in a developing country. Thus there is 'the assumption that nanotechnology R&D, and therefore a potentially active role in global nanotechnology innovation, is limited to the developed countries and beyond the realm of developing countries' (Maclurcan 2009). Yet one might point to South Africa's active role in global partnerships such as ESASTAP and ESASTAP Plus, dedicated platforms for the advancement of global scientific and technological cooperation,[4] or IBSA, an India, Brazil, and South Africa nanotech project working on solar, drug delivery, and nanosponges. Researchers in South Africa are working on a way to incorporate tuberculosis drugs into nanoparticles so they are released slowly into a patient's bloodstream, raising the possibility that daily pills could be replaced with a single weekly dose. Despite the expenses of development, 'the potential advantages of the technology make its pursuit worthwhile. If TB treatment is reduced to a once-a-week dose, the overall costs, both of the drugs and of employing healthcare staff, could be significantly reduced' (Makoni 2010).

Developing countries may still decide to prioritise nanotechnology in terms of budget spend, and global partnerships may go some way to addressing the issue of funding. The question is rather that of whether one should, according to the principle perhaps of distributive justice, argue that a potentially transformative technology such as nano should be more freely available to developing countries.

There is a further issue: that of benefit-sharing. In essence, patenting at the nanoscale could mean monopolising the basic building blocks of life. Whereas biotechnology patents make claims on biological products and processes, nanotechnology patents may literally stake claim to chemical elements, as well as the compounds and the devices that incorporate them. With nanoscale technologies, the issue is not just patents on life – but on all of nature – opening up new avenues for biopiracy (Wetter 2010). Thus Brownsword (2008) has suggested that more attention needs to be paid to how we regulate for benefit-sharing. The European Group on Ethics in Science and New Technologies (EGE) identifies the following ethical questions relating to the development of nanomedicine:

> How should the dignity of people participating in nanomedicine research trials be respected? How can we protect the fundamental rights of citizens that may be exposed to free particles in the environment? How can we promote responsible use of nanomedicine which protects both human health and the environment? And what are the specific ethics issues, such as justice, solidarity and autonomy that have to be considered in this scientific domain? (The European Group on Ethics in Science and New Technologies to the European Commission 2007)

---

[4]The acronym stands for 'European and South African Science and Technology Advancement Program'. See http://www.esastap.org.za/southafrica/bilateral_int.php

The question in broadest terms therefore is one of how developing countries can best benefit from nano, given an economic situation in which they may be globally disadvantaged. The questions for NELSI commentators seem clear, but whether there are such nanoethical processes current in Africa is rather less certain.

If we look to the wider context of bioethics – which has often been seen as a model for nanoethics – we might start by asking whether or not African bioethics actually exists. Murove (2005), arguing that the current discourse on bioethics in Africa is trapped in Western categories of thought and relies heavily on Western analytical philosophy, maintains that an authentic discourse on bioethics in Africa must take cognisance of the fact that most Africans rely on traditional medicine, which often remains the most accessible and affordable system of health for the majority of Africans in rural areas. A UNESCO report gives the figure for sub-Saharan Africa of 85 % of the population as using traditional healers and notes that 'in Ghana, Mali, Nigeria and Zambia, herbal medicines are administered at home as first-aid treatment for 60 % to children with high fever caused by malaria' (IBC 2010).

However, various bodies within and outside Africa have pioneered the movement towards ensuring that medical research in Africa conforms to international ethical guidelines; the Pan-African Bioethics Initiative (PABIN), for example, is a pan-African organisation established in 2001 to foster the development of bioethics in Africa with a particular focus on research ethics.[5] Various ethics workshops and conferences have been held in Africa.[6] Yet apart from some countries in the southern and eastern parts of Africa and a handful of universities in other parts of Africa, there is no formal ethics education in most of Africa's medical schools (Ogundiran 2004). Chadwick and Schuklenk (2004) have questioned the altruism behind training developing world bioethicists in the West and warn against bioethics colonialism. Opportunities to explore and develop local and traditional knowledge to find solutions for the country's health needs can be lost under such paternalism – thus in South Africa there is a move towards taking indigenous knowledge into account, adding value by complementing it with scientific knowledge (and regulating its exploitation) (Motari et al. 2004).

---

[5] Pan-African Bioethics Initiative (PABIN), see http://www.pabin.net/en/index.asp

[6] Workshop on Ethics Review Committees in Africa, Lusaka, Zambia 29–31 January 2001; An International Symposium on Good Ethical Practices in Health Research in Africa, Pan-African Bioethics Initiative Cape Town, South Africa 23–24 February 2001; An International Conference on Good Health Research Practices in Africa. In collaboration with UNDP/World Bank/WHO; Special Programme for Research & Training in Tropical Diseases (TDR/WHO); African Malaria Network Trust (AMANET); Department of Health and Human Services, USA; European Forum for Good Clinical Practice (EFGCP); Institut National de la Santé et de la RechercheMédicale (INSERM), France; and GlaxoSmithKline (GSK- Belgium) Fondation Merieux. 28–30 April 2003 Addis Ababa, Ethiopia; AMANET training workshop on health research ethics in Africa Biotechnology Centre, University of Yaoundé I, Yaounde, Cameroon; Workshop on Ethical Issues in Health Research in Abuja, Nigeria 03–17 December 2001; National Workshop on Ethical Issues in Health Research. Organised by the University of Ibadan, in collaboration with Aids Prevention Initiative Nigeria (Harvard School of Public Health) and Boston University, Harvard: Held at International Institute for Tropical Agriculture (IITA), Ibadan, Nigeria 27 August–01 September 2003

Africa's Science and Technology Consolidated Plan of Action, 2006–2010, is largely a product of NEPAD, which with the African Union (AU) established a high-level African Panel on Biotechnology (APB) to facilitate open and informed regional multi-stakeholder dialogues on, for example, scientific, technical, economic, health, social, ethical, environmental, trade, and intellectual property protection issues associated with or raised by rapid developments in modern biotechnology.[7] In addition, the Nelson Mandela African Institute of Science and Technology (NM-AIST) is a network of S&T institutes across Africa intended to train the next generation of African scientists and engineers, with a view to impacting profoundly on the continent's development through the application of science, engineering, and technology (SET).[8] This Institute aims to stimulate the establishment of science, technology, and innovation courses at postgraduate level in African universities, to build a critical mass of science policy advisors to African governments and the policy sector, and to build and disseminate information and experiences on science, technology, and innovation policy analysis, advice, and development. UNESCO is also helping to set up regional parliamentary fora on science and technology – for example, the Nigerian Parliamentarian Forum on Science and Technology in Abuja in 2006; however there is as yet no regional parliamentary forum on S&T for sub-Saharan Africa (Schneegans and Candau 2007).

Moemeka (1997), noting the 'fundamental African principles of supremacy of the community, value of the individual, sanctity of authority, respect for age and religion as a way of life', argues that one might add the issue of 'democratisation or all-inclusive participation.' Yet at present, South Africa is the only government in southern Africa that has a Science Communication Unit and a Public Understanding of Biotechnology Programme (Mwale 2008). There are however other initiatives such as the science cafes in Kenya (started in 2008)[9] or various conferences in Africa on aspects of risk, technology, and environment. There are various biotechnology associations such as the Agricultural Biotechnology Network in Africa (ABNETA) made up of people from across Africa who want to discuss, support, develop, or use biotechnology in support of agriculture on the continent; AfricaBio, engaged in transferring information about biotechnology and biosafety to all levels of society through information days, workshops, seminars, conferences, exhibitions, websites, newsletters, and technology demonstration (and has an agricultural biotechnology emphasis); and ACTS, a think-tank on the application of science and technology to development that runs fora that discussed emerging new technologies

---

[7] See http://www.nepadst.org/doclibrary/pdfs/doc27_082005.pdf

[8] See http://www.nm-aist.ac.tz/background.html

[9] See http://dougal.union.ic.ac.uk/media/iscience/features/brewing-knowledge/

and issues to do with biotechnology, biosafety, climate change, and the environment.[10] South Africa has an EU partnership on S&T.[11]

Little is specific to nanotechnology, however. In South Africa in April 2006, the Deputy Minister of Science and Technology, Derek Hanekom, launched the South African National Nanotechnology Strategy to support the optimal use of nanotechnology to enhance South Africa's global competitiveness and to achieve social development and economic growth targets. Since 2006, South Africa has set up several centres in the fields of water treatment, health sciences, and energy production. There is also the Nanotechnology Speak2aScientist programme that aims to educate and enhance a public understanding of nanotechnology and stimulate meaningful public debate (Sci-Bono Discovery Centre, Johannesburg).[12]

Kenya and South Africa have adopted national guidelines on bioethics, although, as Langlois (2008) has noted, there are some differences in the 'translation' of the UNESCO Universal Declaration on Human Rights and Bioethics into national principles. The UNESCO article on Cultural Diversity and Pluralism, Article 9, for example, in terms of the Kenyan Guidelines on to gaining informed consent from married women in rural communities, reminds researchers that each of Kenya's 42 tribes will have unique sociocultural backgrounds.

The South African guidelines, in a section on indigenous medical systems, call on researchers to respect the cultures and traditional values of all communities and also state that:

> The challenge to international research ethics is the development of universal rules for research at a time when health care is being delivered within very different health care systems and in a multicultural world in which people live under radically different economic conditions. (Benatar 2001)

The guidelines suggest state-level focus on how researchers should engage with communities and which particular members of society should receive special attention as vulnerable persons. States may need to adopt particular interpretations of the declaration's principles in order to realise them in national and local contexts.

A more 'wholistic' approach to policymaking, with advice taken on local issues, is needed in Africa.

---

[10] SciDev.Net – the Science and Development Network – is a not-for-profit organisation dedicated to providing information about science and technology for the developing world and lists initiatives in nano in sub-Saharan Africa and North Africa (as well as in Asia).

[11] See http://cordis.europa.eu/fetch?CALLER=EN_NEWS&ACTION=D&SESSION=&RCN=24157

[12] See http://www.sci-bono.co.za/home/index.php?option=com_content&view=article&id=94&Itemid=94

References

Acharya T, Darr AS, Dowdeswell E, Singer PA, Thorsteinsdottir H (2004) Genomic and global health: a report of the genomics working group of the science and technology task force of the UN millennium project. University of Toronto Joint Centre for Bioethics, Toronto

Benatar SR (2001) Justice and medical research: a global perspective. Bioethics 15(4):333–340

Brownsword R (2008) Regulating nanomedicine – the smallest of our concerns? NanoEthics 2(1):73–86

Chadwick R, Shuklenk U (2004) Bioethical colonialism? Dev World Bioeth 18(5):i–iv

European and South African Science and Technology Advancement Program. Strengthening Technology, Research and Innovation Cooperation between Europe and South Africa. http://www.esastap.org.za/southafrica/bilateral_int.php. Accessed 14 Jan 2015

European Commission (2005) South Africa and EU aim to promote closer collaboration in research. http://cordis.europa.eu/fetch?CALLER=EN_NEWS&ACTION=D&SESSION=&RCN=24157. Accessed 14 Jan 2015

European Group on Ethics in Science and New Technologies to the European Commission (2007) Opinion on the ethical aspects of nanomedicine. Opinion No. 21. Available via Nanowerk. http://www.nanowerk.com/nanotechnology/reports/reportpdf/report110.pdf. Accessed 14 Jan 2015

Foladori G, Invernizz N, Záyago E (2009) Two dimensions of the ethical problems related to nanotechnology. NanoEthics 3:121–127

Heller J, Peterson C (2007) Nanotechnology: maximizing benefits, minimizing downsides. In de S. Cameron NM, Ellen Mitchell M (eds) Nanoscale, issues and perspectives for the nano century. Wiley, Hoboken, pp 83–96

Hermerén G (2007) Challenges in the evaluation of nanoscale research: ethical aspects. NanoEthics 1:223–237

Hunt G (2008) The global ethics of nanotechnology. In: Hunt G, Mehta M (eds) Nanotechnology. Risk, ethics and law. Earthscan, London, pp 183–195

International Bioethics Committee (IBC) (2010) Draft preliminary report on traditional medicine and its ethical implications. Available via UNESCO. http://unesdoc.unesco.org/images/0018/001895/189592e.pdf. Accessed 21 Dec 2014

Invernizzi N (2008) Nanotechnology for developing countries. Asking the wrong question. In: Banse G, Grunwald A, Hronszky I, Nelson G (eds) Assessing societal implications of converging technological development. Sigma, Berlin, pp 229–239

Jamison A (2009) Can nanotechnology be just? On nanotechnology and the emerging movement for global justice. Nanoethics 3(2):129–136

Langlois A (2008) The UNESCO universal declaration on bioethics and human rights: perspectives from Kenya and South Africa. Health Care Anal 16(1):39–51

Macklin R (2004) Double standards in medical research in developing countries. Cambridge University Press, New York

Maclurcan DC (2009) Southern roles in global nanotechnology innovation: perspectives from Thailand and Australia. NanoEthics 3:137–156

Maclurcan D (2010) A more equitable approach to nano-innovation is needed. Available via Sci Dev Net. http://www.scidev.net/global/health/opinion/a-more-equitable-approach-to-nano-innovation-is-needed.html. Accessed 14 Jan 2015

Makoni M (2010) Case study: South Africa uses nanotech against TB. Available via Sci Dev Net. http://www.scidev.net/global/health/feature/case-study-south-africa-uses-nanotech-against-tb-1.html. Accessed 8 May 2015

Moemeka AA (1997) Communalistic societies: community and self-respect as African values. In: Christians C, Traber M (eds) Communication ethics and universal values. Sage, Thousand Oaks, pp 170–193

Motari M, Quach U, Thorsteinsdóttir H, Martin DK, Daar AS, Singer PA (2004) South Africa – blazing a trail for African biotechnology. Nat Biotechnol 22(December):DC37–DC41

Murove FM (2005) On African bioethics: an exploratory discourse. J Study Relig 18(1):16–36

Mwale PN (2008) Democratization of science and biotechnological development: public debate on GM Maize in South Africa. Afr Dev 33(2):1–22

Nyasani JM (1997) The African psyche. University of Nairobi/Theological Printing Press Ltd., Nairobi

Ogundiran TO (2004) Enhancing the African bioethics initiative. Available via BMC Medical Education 4. http://www.biomedcentral.com/1472-6920/4/21/. Accessed 14 Jan 2015

Pan African Bioethics Initiative (PABIN). http://www.pabin.net/en/index.asp. Accessed 14 Jan 2015

Petryna A, Lakoff A, Kleinman A (eds) (2006) Global pharmaceuticals. Ethics, markets, practices. Duke University Press, Durham

Salamanca-Buentello F, Persad DL, Court EB, Martin DK, Daar AS, Singer PA (2005) Nanotechnology and the developing world. PLOS Med 2(5):e97. doi:10.1371/journal. pmed.0020097

Schneegans S, Candau A (2007) Science in Africa: UNESCO's contribution to Africa's plan for science and technology to 2010. UNESCO, Paris

Schummer J (2007) The impact of nanotechnologies on developing countries. In: Allhoff F, Lin P, Moor J, Weckert J (eds) Nanoethics: the ethical and social implications of nanotechnology. Wiley, Hoboken, pp 291–307

University World News (2010) Africa. Nineteen countries pledge to promote science. http://www.universityworldnews.com/article.php?story=20100911201707964. Accessed 14 Jan 2015

Valdivia WD (2011) Innovation, growth and Inequality: plausible scenarios of wage disparities in a world with nanotechnologies. In Nanotechnology and the challenges of equity, equality and development, Yearbook of nanotechnology in society. Springer, Dordrecht, pp 145–164

Wetter KJ (2010) Big continent and tiny technology: nanotechnology and Africa. Foreign Policy in Focus, Washington, DC, 15 October 2010

Printed in the United States
By Bookmasters